Max Wertheimer
Productive Thinking

Birkhäuser

Classic Texts in the Sciences

Series Editors
Jürgen Jost
Armin Stock

Classic Texts in the Sciences offers essential readings for anyone interested in the origin and roots of our present-day culture. Considering the fact that the sciences have significantly shaped our contemporary world view, this series not only provides the original texts but also extensive historical as well as scientific commentary, linking the classical texts to current developments. *Classic Texts in the Sciences* presents classic texts and their authors not only for specialists but for anyone interested in the background and the various facets of our civilization.

More information about this series at http://www.springer.com/series/11828

Max Wertheimer

Max Wertheimer
Productive Thinking

Viktor Sarris
Editor

Birkhäuser

Author
Max Wertheimer
New York, NY
USA

Editor
Viktor Sarris
Frankfurt am Main, Germany

ISSN 2365-9963 ISSN 2365-9971 (electronic)
Classic Texts in the Sciences
ISBN 978-3-030-36062-7 ISBN 978-3-030-36063-4 (eBook)
https://doi.org/10.1007/978-3-030-36063-4

Originally published in German language: Max Wertheimer, Produktives Denken by Viktor Sarris

This book is published under the imprint Birkhäuser, www.birkhauser-science.com, by the registered company Springer Nature Switzerland AG.
The registered company address is: Gewerbestrasse 11, 6330 Cham, Switzerland

Max Wertheimer *(1880–1943), the main founder of Gestalt psychology, taught and conducted research in perception and psychology of thinking in Frankfurt am Main and in Berlin. He emigrated to the USA in 1933 to continue his research and teaching at the New School for Social Research (New York). His work* Productive Thinking *(1945) was posthumously published there as his chief work.*

Viktor Sarris *is Professor Emeritus of Psychology, in particular general psychology, and the former holder of the Max-Wertheimer Chair at the Goethe University in Frankfurt. His work focuses on perceptual as well as cognitive psychology, comparative psychophysics, and the history of psychology.*

Preface for the New Edition

Max Wertheimer (1880–1943), a pioneer of psychology in the last century, has promoted the development of cognitive science to a large extent, especially in the psychology of perception and thinking. His book *Productive Thinking*, written up in New York during his exile years, was published 75 years ago, and it became one of the milestones of the booming research on "creativity." Consisting of many examples for goal-oriented creative processes in thinking and problem solving, this work - starting from simple geometrical tasks as provided for elementary-school children via social psychological issues in conflict resolutions toward the development of the theory of relativity by Albert Einstein - leads through a multifaceted and thought-provoking network of creative thinking. Today, Wertheimer's treatise is all the more relevant because many of the modern computer and internet facilities threaten our children's creative abilities.

At the center of the book are the concepts of insight and understanding in the human thought processes as involved in creative problem solving ("seeing the light") - when "the scales fall from your eyes," so to speak. Wertheimer's earlier publications in Germany constitute the basis for his book on productive thinking. However, since his idiosyncratic style of writing is often not easily to be followed, it is felt that an introductory commentary is needed, thus facilitating the understanding and appreciation of the complex lines especially of the gestalt-theoretical argumentations. Recent summaries of the book have been provided elsewhere in German (Sarris, 1996; Sarris & Michael Wertheimer, 2018).

This *new edition* is based on Wertheimer's original first edition (1945); see the respective note in the end of the following *Introduction* ("Reception of the book," p. 18). The German version of this new edition has been recently published by Springer (Heidelberg, 2019).

I am grateful for the support of my introductory comments as provided by several colleagues, especially Lothar Spillmann, Freiburg; Armin Stock, Würzburg; Michael Wertheimer, Boulder, CO; and Friedrich Wilkening, Zürich. For the *new edition* Michael

Wertheimer, the son of Max Wertheimer, has translated his father's vita from the German original into English. The style of my English text has been greatly improved with the help of Susan Opotow, New York. I thank them all.

Johann Wolfgang Goethe University
Frankfurt am Main, Germany
September 10, 2019

Viktor Sarris

Contents

Introduction to Max Wertheimer: *Productive Thinking* 1

Viktor Sarris

There are entities where the behavior of the whole cannot be derived from its individual elements nor from the way these elements fit together; rather the opposite is true: the properties of any of the parts are determined by the intrinsic laws of the whole.

Max Wertheimer (1925), Lecture at the KANT Society on 17 December, 1924.

Brief Evaluation of the Book

The above motto is the main credo of the gestalt-theoretical approach in psychology (Westheimer, 1999), and is taken from Wertheimer's original article which was published 20 years in advance of his book "Productive Thinking." The book, which appeared posthumously in his New York exile (1945), advances the idea that

- the contemporary logical and association-theoretical approaches were not able to explain the crucial characteristics of productive thinking and learning, especially its main features of understanding and insight ("seeing the light");
- in contrast, the gestalt-theoretical perspective provides a reasonable basis for such cognitive issues and questions which the former traditional theories had avoided if not completely ignored;

V. Sarris (✉)
Frankfurt am Main, Germany

M. Wertheimer, *Max Wertheimer Productive Thinking*, Classic Texts in the Sciences,
https://doi.org/10.1007/978-3-030-36063-4_1

– Wertheimer's account is still challenging the modern theories or models of cognition which are leaving the central issues of productive thinking unexplained until today.

The book has a long history by itself (see below, "History of the book"). Wertheimer's life and work, often being misrepresented in former published accounts, has been treated in length only a few years ago (Ash, 1995; King & Michael Wertheimer, 2005).

Biography of Max Wertheimer

Max Wertheimer was born in Prague, in the former Austria-Hungary, 15 April, 1880 only a few feet away from the Old-Synagogue which was the cultural and religious center of Old Prague. His father Wilhelm (1853–1930) was a successful business professor of his own *Handelsschule* (college of commerce) in Prague, and he published several books on accounting which were standard texts in those days. His mother Rosa, nee Zwicker (1855–1919) was well educated in German literature and music; she was a passionate amateur violonist and had great influence on Max's cultural education. After his highschool graduation, with Latin and Greek included, Max studied law, physiology, philosophy, and psychology first at Charles University of Prague, later at the universities of Berlin and Würzburg. His dissertation in legal psychology became finished under Oswald Külpe's direction. His chronology is contained in Table 1, and it includes some of his major publications.

Wertheimer who had started his dissertation already in Prague, continued his graduate studies in Berlin and Würzburg. He wrote his *Vita* for Würzburg's Faculty as follows (1904; cf. Wertheimer, 1905):

> I, M a x W e r t h e i m e r, Austrian citizen, was born April 15, 1880 in Prague (Bohemia), son of business school director Wilhelm Wertheimer and his wife Rosa (nee Zwicker), of Jewish faith. After completion of the Piarist elementary school, the Prague new-city highschool, and five semesters at the faculty of the Prague law school, during which I simultaneously attended philosophical lectures and seminars, I was enrolled from spring 1901 for three semesters at the philosophical faculty in Berlin, (and) since then at the philosophical faculty of Würzburg. I enjoyed instruction, assistance, and support in my studies in the field of experimental psychology at the psychological institutes of Berlin and Würzburg and at the physiological institute in Prague, especially from Messrs. Külpe, Stumpf, Schumann, (Johannes) Gad (. . .). I obtained friendly assistance first from Professor Hans Gross (Prague), then from Professor (Friedrich) Schumann (Berlin), in preliminary experiments in the field of the present work. I owe special thanks primarily to Profesor O(swald) Külpe (Würzburg) for abundant assistance, instruction, and support by word and deed. Gratitude should also be expressed once more (. . .) to the participants in the experiments.—(Source: Sarris & Michael Wertheimer, 1987; translation from the original into English by Professor Michael Wertheimer, June 12, 2019).

Wertheimer's intellectual and cultural interests were very broad already during his student's career (Michael Wertheimer, 2014). Until today it is not quite clear why he

Table 1 Max Wertheimer, 1880–1943: Chronology

Chronology: Max Wertheimer (1880–1943)	
1880	Born in Prague on April 15, 1880
1898–1903	University studies in Prague and Berlin
1904	Dissertation thesis with O. Külpe at Würzburg, about a topic on forensic psychodiagnostics
1910	Research worker at the Institute of Psychology at Frankfurt/Main. Beginning of the experiments on stroboscopic motion ("phi phenomenon"). W. Köhler and K. Koffka Schumann's assistants and Wertheimer's main subjects (life-long cooperation from that time onwards)
1912	First epoch-making publication on the phi phenomenon: "Experimentelle Studien über das Sehen von Bewegung" (*Experimental Studies on the Seeing of Motion*)
1912–1916	Habilitation thesis (1912), Lecturer (*Privatdozent*) at Frankfurt/Main
1918–1929	Lecturer (1918) and Associate Professor (1922) at Berlin University
1921	Foundation of the journal *Psychologische Forschung,* together with W. Köhler, K. Koffka, and others (most important journal of Gestalt psychology until 1938)
1929–1933	Full professorship of psychology ("Philosophie, insbesondere Psychologie") at the psychology institute of the Faculty of Sciences at Frankfurt's J.W. Goethe University (predecessors K. Marbe 1904–1909, F. Schumann 1910–1929)
1933	Employment ban by the Nazi government; professor at the New School for Social Research in New York, founder of its Department of Psychology
1934–1943	Co-editor of *Social Research;* active in rescue efforts and, since 1938, member of the Committee for Displaced Foreign Psychologists, of the American Psychological Association (APA)
1943	Died in New Rochelle, NY, on October 12, 1943
1945–1959	Posthumous treatises: *Productive Thinking* (1945), *On Discrimination Experiments* (1959)

Source: Sarris, 1989

moved as a student from Prague to Berlin and then to Würzburg, but there is enough evidence that he collected the main experimental data for his dissertation at all three universities; it is also known that he took most of his psychology courses at the latter two institutions (King & Michael Wertheimer, 2005). In fact, throughout his student years, he had studied under the guidance of renowned professors. Interestingly, one of his additional teachers in Prague had been Christian von Ehrenfels (1890), the most important predecessor of gestalt psychology. Also influential for Wertheimer's later general thinking were his following friends in his early life—namely: Before, during, and after his student career at Prague as well as in Berlin and Frankfurt, Wertheimer had a couple of close friends, some of whom became famous by themselves—i.e. in Prague especially the writer Max Born and the ethno-historian and psychologist Johannes Urzidil, in Berlin the music-ethnologist and psycho-acoustician Erich von Hornbostel, the sculptor Käthe Kollwitz and the physicist Albert Einstein, in Frankfurt the neurologist and psychiatrist Kurt Goldstein, the psychologist Adhemar Gelb, and the theologist and philosopher Paul Tillich (Sarris, 1997).

After his dissertation Wertheimer continued his basic research interests, and in 1912 he wrote up his second main thesis ("Habilitation") at Frankfurt on apparent motion perception under the direction of Friedrich Schumann (Wertheimer, 1912b: *phi phenomenon*; i.e. an important type of motion illusion generated by stationary stimuli). Especially with this monographic paper he soon became an internationaly wellknown scholar (Sarris, 1987, 1989, 1995; Ash, 1989, 1995; cf. also Sekuler, 1996; Westheimer, 1999; Spillmann, 2012; Gundlach, 2014; Stock, 2014). Already before, and also thereafter, he engaged in a series of further gestalt-theoretical topics such as some music psychology issues (Wertheimer, 1910) and special ethno-linguistic problems in thinking (Wertheimer, 1912a). In addition he had become interested in the study of the symptoms of aphasia in brain-injured patients (Wertheimer, 1923; cf. Sarris & Michael Wertheimer, 2001). Clearly, Wertheimer's university career became shaped mostly by his extensive work on motion perception (Wertheimer, 1912b) which is known as the beginning of the gestalt school in Frankfurt am Main (Sarris, 1987; Ash, 1989).

After Max Wertheimer's stay in Berlin for more than 10 years, mostly because of WW I reasons, he became full professor and institute director of Frankfurt's institute of psychology in 1929, as the successor of his former mentor Friedrich Schumann. There he continued his experimental research on perception and cognition—until its sudden halt: In March 1933 Wertheimer fled with his family to Marienbad, the former Czechoslowakia, for political reasons; from there he emigrated into the USA in September of the same year. Other gestalt psychologists also emigrated to the USA, sooner or later (Wolfgang Köhler, Karl Duncker, Kurt Lewin, and also Rudolf Arnheim, George Katona, Erwin Levy). In New York he was teaching at the New School for Social Research, which became famous under its name "University in Exile." In the USA he avoided the German language, i.e. he published only in English like the present book and also his articles on humanistic-psychological problems on truth, ethics, freedom, and democracy (Wertheimer, 1934, 1935, 1937, 1940; cf. Albert Einstein's Preface to it, 1950/1991). With no laboratory at the New School, Wertheimer no longer conducted psychological experiments, as this is illustrated by the assertion of Mary Henle, a leading historian of psychology:

> It is impossible to compare the facilities at Frankfurt and at the New School. The New School's University in Exile, its first graduate faculty, started in 1933 as a rescue operation for displaced German scholars. It had no facilities. It was not yet giving degrees, so the student body was very diverse; by no means all of them prepared to do graduate work. Wertheimer knew practically no English; I believe his first seminar was given in German. The financial difficulties of the New School were famous. I could go on, but the remarkable thing is that . . . Wertheimer attracted to his seminars such psychologists as S(olomon) E. Asch and others. The whole operation was a struggle, with no assurance that it would last. It is a miracle that it did.—(Professor Mary Henle, former Wertheimer student at the New School, in her letter to V. S., October 23, 1986; quoted from Sarris, 1996, 2012.)

Hardly known today by psychologists and other scholars in Europe and in the USA are Wertheimer's extended rescue activities for endangered scientists during this dark period of history. These activities must have taken much time and energy from him. His former guest student in Frankfurt, Edward B. Newman, who became later a professor at Harvard, remembers Wertheimer's humanitarian work as follows:

> (Wertheimer's) other activity which I recall clearly had to do with the flow of refugees through his house in New Rochelle. Mostly, these were scholars who had just arrived in New York. Just why they came, how they knew where to call, and what they may have been their destination is quite unclear. They came at every time of day. Michael, Max Wertheimer's son, remarked to me that they often stayed for a modest supper, and I can confirm that. And their number was legion. As a guess, I would say that upwards of 250 refugees made the pilgrimage to New Rochelle in the years 1933 to 1937. So far as I can discover, this was a purely informal network.—(Edward B. Newman, 1989; quoted from Sarris, 1989, 1997.)

Wertheimer wrote his *opus magnus* during these years; he finished his book manuscript only a few weeks before his death in October 1943. As noted above, he had prepared important preliminary work already in Germany, but his work remained unfinished (see below, "Wertheimer's work from a modern perspective").

History of the Book

Wertheimer's first publication from a gestalt perspective was his paper "Über das Denken der Naturvölker" (Wertheimer, 1912a) with its emphasis on number comprehension in indigenous people. Before that article he had already published his music-psychological paper containing gestalt-theoretical ideas (Wertheimer, 1910). But of even higher significance for the "Productive Thinking" book became his monograph "Über Schlußprozesse im produktiven Denken," which he wrote during World War I in Berlin but published only in 1920—with some fundamental illustrations of creative thought processes (Wertheimer, 1920). Wertheimer was also teaching these topics in his university courses in Berlin and Frankfurt. He offered his perception and cognition lectures first in Germany, then in the USA (Fig. 1).

As in Frankfurt, Wertheimer demonstrated some of his major "gestalt" examples also in the New School's lecture hall on the piano. In New York he wrote his book manuscript by the language assistance of some of his main students and collegues such as Rudolf Arnheim and Solomon E. Asch. The book became co-edited posthumeously by his friend and collegue Wolfgang Köhler, among others (cf. here, pp. 32 & 34). All in all, the history of Wertheimer's book took about 35 years. It should be also mentioned that Karl Duncker, a former Wertheimer-Köhler student in Berlin, had published his own gestalt-theoretical book on productive thinking 10 years ahead (Duncker, 1935).

Fig. 1 Max Wertheimer at the New School for Social Research, New York: pencil drawing by the Wertheimer student Rudolf Arnheim, May 20, 1942. Source: Arnheim, 1989

Main Topics of the Book

The book consists of seven chapters, which contain a host of lengthy examples illustrating quite different kinds of "productive" thinking processes. It is written in a teaching style which encourages the reader to grasp and reflect on Wertheimer's major issues of creative problem solving. The book's chapter headings are:

- The area of the parallelogram
- The problem of the vertical angles
- The famous story of young Gauss
- Two boys play badminton; a girl describes her office
- Finding the sum of the angles of a polygon
- A discovery of Galileo
- Einstein: the thinking that led to the theory of relativity

These chapters deal with the calculation of the area of the parallelogram; the problem of the equality of vertical angles; the discovery of the young Carl Friedrich Gauss with his simple method of finding the sum of a regular algebraic series; how two boys find a creative solution in favor of a fair style of badminton playing, and how an inexperienced (frustrated) secretary might reorganize her office work; Wertheimer's own finding re the calculation of the angles of a three-dimensional polygon; how Galileo Galilei discovered the physical law of inertia; and, finally, how Albert Einstein's thinking led to the theory of relativity.

In everyday life the non-creative, i.e. unproductive thoughts and activities dominate. As compared to that, Wertheimer's **introductory chapter** (pp. 36–48) starts with the following questions:

> What occurs when, now and then, thinking really works productively? What happens when, now and then, thinking forges ahead? What is really going on in such a process? (…). Have you been serious about what this thing called thinking may be? (…) What happens if one really thinks, and thinks productively? (…) And in connection with all these questions: how improve thinking? your thinking? thinking itself?—(Wertheimer, pp. 36f)

Wertheimer underlines his claim that the messages as provided by the traditional socalled association theory of Wilhelm Wundt (1832–1920) in Germany and the behaviorism of Edward L. Thorndike (1874–1949) in North America have blurred, if not simply ignored, the factual psychological problems as involved in creative thinking. In stark opposition stands Wertheimer's gestalt-theoretical position which, however, was not known or at best only superficially grasped by most of the contemporary psychologists in the USA.

The Area of the Parallelogram (Chap. I)

This chapter, the longest one in the book, consists of four parts (part I to IV). It is divided into 40 sections and deals with the calculation of the area of the parallelogram by school children in New York City (pp. 49–113). It is guided by the psychologically as well as pedagogically motivated intentions regarding a "productive" thinking attitude, i.e. to illustrate reasonable versus routine learning at school ("insight" as opposed to "school-drill"). Accordingly, the treatment of the insightful steps, as opposed to mere trial and error messing abouts, follows Wertheimer's aim to work out the socalled real thought processing (interactions between the teacher and the pupils in the class as guided by Wertheimer as the psychologist). Two typical opposing findings with many pupils of elementary schools located in New York City are the following ones:

> There are extreme cases of thoughtless reactions, in which a child, if simply given the figure (the parallelogram, Fig. 6), repeats what the teacher has said word by word like a blind slave (*sic*) and mumbles: "One perpendicular from the upper left corner," and draws it; then, "Another from the upper right corner," and draws it. (…) On the other hand, it *does* happen that children even as young as six, who have learned nothing of geometry, after being shown

> briefly of how to find the area of the rectangle, *do* produce the solution for the parallelogram in fine, original processes without being taught what to do.—(Pp. 52f; emphasis: in the original; cf. Wertheimer's alternative expression for *insight*: "seeing the light", pp. 102f)

In general, the children had to find out, with the help of auxiliary drawings, that a given parallelgram must first be reduced to the area of the corresponding rectangle a x b in order to determine the exact area of a given paralellogram (cf. §23, pp. 73ff; see also fn 6, p. 74 with the reference to a report with a "... solution of a very similar problem which ... does lead to a more direct understanding, see Appendix I", pp. 252–256). Sometimes Wertheimer's task led to: "... no reaction at all. Or someone says, *Whew! Mathematics!* and dismisses the problem with, *I don't like mathematics*" (p. 80).—However, there were also clear examples of "insight" and "understanding" of the problem at hand; this is illustrated by the following case:

> Given the parallelogram problem (to a 51/2-year-old child), after she had been shown briefly how to get at the area of the rectangle, she said, "I certainly don't know to do that." Then after a moment of silence: "This is *no* good here, "pointing to the region at the left end; "and *no good here*," pointing to the region at the right (Fig. 27): "It's troublesome, here and there." Hesitatingly she said: "I could it make right here ... but ..." Suddenly she cried out, "May I have a (pair of) scissors? What is bad *there* is just what is needed *here*. It fits." She took the scissors, cut the figure vertically, and placed the left end at the right.—(P. 83; emphasis in the original text.)

One of the main questions for Wertheimer's analyses was: Which factors are most relevant for the phenomena of perceptual "past experience" and "transposition" as inherently related to productive thinking? The role of past experience (familiarity)—this was important for him—must *not* be downplayed as long as true creative thinking will be promoted; namely:

> There are still psychologists who, in a basic misunderstanding, think that gestalt theory tends to underestimate the role of past experience. Gestalt theory tries to differentiate between and-summative aggregates, on the one hand, and gestalten, structures, on the other, both in sub-wholes and in the total field, and to develop appropriate scientific tools for investigating the latter. (However), it opposes the dogmatic application to all cases of what is adequate only for piecemeal aggregates. (...) Past experience has to be considered thorougly, but it is ambiguous in itsellf; so long as it is taken in piecemeal, blind (i.e. routine: V.S.) terms it is not the magic key to solve all problems.—(P. 100)

The creative problem solving processing is illustrated by means of three structural schemata (pp. 86ff). Of great interest are also Wertheimer's comments on the perceptual basis of the "transpositional" aspects of the genuine problem-solving processes involved (pp. 102f). In sum, the chapter underlines Wertheimer's position by a comparison between the "traditional logic" and the perspective of "gestalt theory"; and it ends with some unique rhetorical appeal to the reader: "Think it over yourself, reader. Compare this assertion, in the spirit of traditional logic, with a genuine finding process. Maybe you will agree (with me); maybe not" (p. 106; cf. also pp. 111ff).

The Problem of the Vertical Angles (Chap. II)

This chapter, which consists of four parts (part I to IV), deals with the mathematical proof of the equality of vertical angles (pp. 114–123). An elemenary school teacher may offer a task in geometrics as follows: "Two straight lines intersect and form two angles, a and b. Can you prove that they are equal?" (p. 114, cf. Fig. 43). Some of the smart pupils may be surprised about this seemingly trivial task, and they ask why they should bother:" "What are you asking? Isn't it obvious? Naturally the(se) angles are equal; can't anyone see it?" (p. 114). In fact, the teacher must first "waken up" the children's problem-solving attitude since often the demand for a mathematical proof is not immediately understood. If a pupil merely repeats "word for word" what the teacher has taught him before, he or she may be at a loss to understand: "Does he repeat what he heard *blindly*, like a slave (sic), or has he grasped, understood?" (p. 115: emphasis: V.S.). Indeed, several children were lost at this point:

> There are pupils who are seriously disturbed if the teacher (. . .) uses a diagram with other than the habitual labels. This is no proof of the assertion that "mind is ruled by habits throughout." It is (only the) proof that the particular individuals stick blindly (i.e. repetitively: V.S.) to "what they were taught."—(Pp. 116f).

Wertheimer asks which may be the explanatory basis for this repetitive "sticking" to the familiar knowledge of many pupils? He argues that mostly the influential American behaviorist and learning theorist Edward L. Thorndike was responsible for this mental "blindness" in learning and memory. In stark contrast, he proposes a psychologically and mathematically sophisticated strategy in that the vertical angle problem is understood as "a part within a given whole," as this is further exemplified by the help of two auxiliary "structural" schemata (cf. Fig. 53, p. 120 and Fig. 58, p. 122). Interestingly, in the end of the chapter Wertheimer takes a critical look also at Immanuel Kant's philosophy of science according to which the insuffient habit of the socalled "successive" (routine) teaching mode is over-emphasized in school learning; indeed, he is convinced that this is due to Kant's erroneous assumption about the general nature of human thinking; in fact, it is hold that in line with Kant's traditional logic the education has been hampered until today because it followed this "practice of formal logic in and-sums of data" (p. 123; cf. also fn 4).

All in all, the chapter vividly illustrates Wertheimer's didactical purpose to demonstrate the contrast between the children's "blind" attitude (school drill) and the gradual awakening of a "productive" understanding in mathematics.

The Famous Story of Young Gauss (Chap. III)

This chapter, which is divided into five parts (parts I to V; pp. 124–158), discusses the solution of the 6 year old boy Carl Friedrich Gauss (1755–1855). His teacher gave the following test to the class and asked: "Which of you will be the first to get the sum of 1 + 2 + 3 + 4 + 5 + 6 + 7 + 8 + 9 + 10?" (p. 125). The genius Carl Friedrich, with his immediate solution ("55"), bet all his classmates high: "How the devil did you get it so quickly?" asked the surprised teacher. Well, young Gauss had made intuitive use of an organizational principle of finding the respective sum, namely he had calculated the single additions of $1 + 10 = 11, 2 + 9 = 11, 3 + 8 = 11, 4 + 7 = 11$, and $5 + 6 = 11$ which resulted, of course, in $5 \times 11 = 55$ (cf. Fig. 60, p. 125; cf. also Fig. 59, p. 124). In fact, as he had immediately recognized, there were 5 pairs of addenda ("11") of the given mathematical series (therefore, 5 times 11 amounts to 55).

In so doing young Gauss discovered an important mathematical rule, i.e. the summation formula $S = (n + 1)/(n/2)$ whereas n is the sum of the initial numbers of a mathematical series. This means psychologically: "Essentially there is involved the regrouping, the reorganization of the series in the light of the problem. This is no *blind* (i.e. routine: V. S.) regrouping; it comes about reasonably as the subject seeks to grasp the inner relation between the sum of the series and its structure. In the process the various items clearly gain a new meaning . . ." (p. 126; emphasis: V.S.). In his replicational investigation, Wertheimer offered the same Gauss task to many pupils of various ages ". . . to see whether a good solution would be found, and what helps, what conditions might bring it about. In order to study the steps and the features involved, I employed *systematic variations*" (p. 125; emphasis: V.S.). He suggested three types of acceptable solutions, for which it was necessary to understand that an increase of the numbers of a given series from left to right corresponds to a decrease from right to left (". . . there comes a sudden discovery"; cf. p. 92). In his considerations Wertheimer admits that certainly not all of the participants were able to solve the problem—indeed, in the beginning he himself had some difficulties with this task:

> I confess that I myself was for a long time at a loss to see how one could reasonably have come to the idea of doubling. It had looked to me like a trick, as it does to many, like a chance discovery. (. . .) But the moment one tries to get at the psychological process in productive thinking, one has to investigate, to view the terms in their functional (gestalt-like) meanings.
>
> These bring about the solution in the sensible, productive processes; they constitute the basic difference between finding the formula in a sensible way and finding it by blind learning or by chance trial and error.—(P. 130; emphasis: V.S.)

Whatever the individual child's thought processing may have been in its details, his findings led Wertheimer to the conclusion that for problem solving its "structural" basis must first be "seen" and recognized as such; namely: ". . . first the problem is seen, realized. It calls for grasping the concrete structure of the series in the light of the problem. The aim of discovering the inner relation between structure and task leads to regrouping, to structural reunderstanding. The steps and operations . . . are performed in view of the whole situation, of the functional need for them, not by blind accident nor as the thoughtless repetition of an old rule-of-thumb connection" (p. 131; cf. also fn 7, p. 131). Thereby the issue of "functional fixation," in its relation to perceptual familarity (see fn 11, p. 141), is stressed: "An investigation of the blinding effects of mechanical repetition in sequences of assigned tasks was begun in the Berlin Institute in 1924, (Karl) Duncker and (Karl) Zener obtained striking results. In recent years (c. 1940–1942), a student of mine, A.(braham) Luchins, has made a comprehensive investigation in schools of this effect. (. . .) It is striking how easy it is, through some kind of mechanization, to *blind* even bright, highly educated subjects with the beloved (sic) method of repetition. (. . .) Indeed, (Abraham) Luchins and (Solomon) Asch made (extensive) experiments concerning such theoretical issues (pp. 148; emphasis & addition: V.S.). Interestingly, Wertheimer discussed all these elementary-school findings with teachers, mathematicians, logicians, and also with other psychologists in the light of the gestalt-theoretical meaning re the functional "whole" characteristics in problem solving. In the end, also the somewhat ambiguous issue of mathematical "exactness" is treated along with the mentioning of the empirical studies of other Wertheimer students in New York City like Catherine Stern and George Katona (e.g., pp. 148f; see also part V, pp. 150ff).

Two Boys Play Badminton; a Girl Describes Her Office (Chap. IV)

This chapter, divided into two parts, has a social-psychological as well as educational orientation (Part I & II, pp. 159–182). It is shown that goal-directed "productive" solutions may be obtained in many social settings. In his first example Wertheimer discusses a badminton game between two boys A (12 yrs) and B (10 yrs), which however became soon very boring because of the older boy's much better performance. Finally the boys agreed in jointly finding a simple rule-changing solution of the game, according to A's insight ("seeing the light"): "I have an idea—let us now play this way: let's see how long we can keep the bird (ball) going between us and count how many times it goes back and forth without falling. (. . .). He spoke happily, like someone who had discovered something. It was a new (good) thing for him (. . .). B agreed happily" (p. 166: addition: V.S.). Wertheimer uses this simple rule-change example as an illustration of the general need for creating *co-operational* games, e.g. also in playing chess:

> I know chess players who are intensely disturbed if fine constellations in a game were disrupted (. . .). Why? They like a good game; they have no interest in winning by some silly mistake of their opponents. (. . .) In general I think there should be more games of the co-operative, creative sort instead of merely competitive ones.—(P. 166, fn 10)

Wertheimer asserts also from a philosophical point of view that the individual attitude—not only the individual style in badminton or chess playing—becomes most important for any problem solving concerning social issues. He makes this point by the aid of some graphic schemata which describe the social ("relational") network at hand (cf. Fig. 92 through Fig. 95, p. 167; cf. also p. 169). From his analysis he draws the following conclusion: "The role of the merely subjective interests of the *self* (ʹegoʹ) is, I think, much overestimated in human actions. Real thinkers forget about themselves in thinking .... This (game) was only a modest little story. Yet some of its characteristics as, for example, the essential progress which occurred through a change of centering, are, I think, characteristic of profoundly important accomplishments in human beings, in *human society*" (p. 170; emphasis: V.S.).

Wertheimer's second example, which looks like a trivial everyday social issue, deals with a young inexperienced, probably frustrated, office girl who had difficulties to describe clearly her social status in her team. As a matter of fact, she seemed "to relate all things (of her office) to herself, as if she were always the center of the given situation (...); she had not made the real picture clear ... but (enumerated the social interrelations) in a bewildering succession" (p 172). One wonders what this example may tell the reader; here is Wertheimer's statement in its essence, namely:

> This is an inncocent example of a foolish attitude in life and in thinking, which (however) often has considerable consequences for shaping one's views and one's actions. (...) When *self-centering ("ego")* is extreme, it becomes a well-known symptom of a *psychopathological* (sic) state that often leads to a precarious situation both in social and in personal affairs.—(Pp. 172f; emphasis: V.S.)

How may the reader interpret the present case regarding this clerk's "foolish attitude"? Wertheimer argues that she may have described her office situation in a "correct" but in an incomplete and *bewildering* way: "The girl had stated all the (single) relations involved, but in a rather (piecemeal) disturbing, miscentered aggregate" (p. 172). Theoretically, there are four different types of possible "descriptions" for these social interrelationships at hand, but only one is judged to be truly adequate, i.e. the "fourth description" as outlined by Wertheimer (cf. Figs. 96 – 113, pp. 173–177). In fact, for the given "relational network" between the office individuals—the boss, two secretaries, four clerks—it may be said: "Viewed piecemeal, the first and second descriptions are in a way correct, and complete in all details, but they introduce a centering, a grouping that is blind to, and in violation of, the logical hiearchy in the (real) situation. They put things which are different in nature and let things appear as structurally different which are not. (...) Her description distorts the (real whole) structure; it fails to do justice to the structural (valid) meanings of the parts" (p. 178). In the end of the chapter, one does not know for sure if the office girl did *re*-adjust her personal cognitive "distortions" of her office situation; however, it seems that her state of affairs "... gets some vague idea ... which involves directions of recentering, until suddenly the picture crystallizes into the complete new (valid) structure" (p. 182).

The concepts of "attitude" and "functional fixation" in both examples are all the more important as they illustrate the everyday situation of "not seeing the beam in your eye";

indeed, they remind the reader of the key notions of "prejudices" and "anchoring" effects in the modern literature on social cognition (cf. for instance Mussweiler, 2003).

Finding the Sum of the Angles of a Polygon (Chap. V)

This chapter deals with another algebraic question concerning Max Wertheimer's *own* creativity in goal-oriented problem solving (pp. 183–194). This was the initial task situation: The sums of all angles for different geometric figures such as triangles, rectangles, hexagons, and still other polygons are different from each other—for a triangle the sum of its angles is 180 degrees, for rectangles it is 360 degrees, for the hexogon it's 720 degrees; and so on (cf. Fig. 122, p. 183). Surprisingly, one of Wertheimer's friends in a gathering with his collegues—an artist—asserted: "The sum of the angles of all such figures must of course be the *same* in every case" (p. 183). When all the others laughed and mocked, only Wertheimer felt that the artist's claim might be "right" somehow. And on the same day he began to think about its underlying mathematical basis, however first without any clear idea of its real meaning. But this self-imposed task led Wertheimer to some longterm thought processing—here is an example for it (cf. Fig. 123 through Fig. 130; pp. 184ff):

> The following day … this idea came suddenly to me, vague, indefinite and uncertain (…). Three days passed. Whatever I did, there was always present this same strong feeling, the feeling of something *unfinished*, of being directed toward something I could not grasp … I could not formulate it concretely (…). Again after two days the following question arose: "If I have the point, there is a complete angle around it. If I have *a straight line*, there is also an angular space around it. …"—(Pp. 184f; emphasis in the original)

A few more days passed without any task-solution; but during this still unfinished solution phase something happened in Wertheimer's mind, namely:

> As the solution came clear, requirement became urgent: If this line of thought really goes to the heart of the matter, then it should also hold for any closed figure. It should hold for three-dimensional solids, for four-dimensional, n-dimensional, indeed for all closed figures. (…) In *6 weeks* of hard work I succeeded in actually understanding (at least the) three-dimensional figures. (…) But the process (still) went right on. After a great deal of effort, the solution for three-dimensional solids came in a *half-wakening* moment during the night. Although I could not recall having written anything, in the morning I found the … (correct) formula on a slip of paper.—(Pp. 186–191; emphasis: V.S.; cf. FIG. 130 through FIG. 139)

However, Wertheimer's task was not completely finished yet (Fig. 140, p. 192): "The following days were devoted to exact demonstrations (and calculations), including spheres, etc. …(But) I shall stop here in the report, with the *joyful moment* in which the inner relation between closure and the sum of angles became translucid in solids as in plane figures" (p. 192; emphasis: V.S.). In conclusion, reading Wertheimer's lines of thinking may please also today's readers, perhaps especially in respect with the additional auxiliary notes as provided in the text (cf. fn 4, p. 191; see also the summary in eleven points, pp. 192–194).

A Discovery by Galileo (Chap. VI)

This chapter, as divided into three parts, is the shortest in the book (part I–III; pp. 195–202). In contrast to the everyday school-life tasks, Galilei Galileo's (1564–1662) discovery of the law of physical bodies occupies an important place in the history of science. Indeed, "... one of the great features in Galileo's thinking was his ability to get ... a clear, clean structural insight against the highly complex and sophisticated background of the motion velocity acceleration" (pp. 195f). In part I the initial situation of motion physics (inertia law) is sketched, in three acts (pp. 196f, cf. Fig. 142); then, in part II (pp. 198f, Fig. 143 through Fig. 145) and in part III (pp. 199f), Galileo's thoughts are described that led to a major breakthrough re the law of inertia (pp. 198–200).

In part II the distinction is made between the "decrease of negative acceleration (retardation)" and "positive acceleration": "When a body is thrown up, we have not positive but negative acceleration. The body slows down in the course of its upward movement ... (p. 198). What, then, are the further thought processes when Galileo suddenly noticed a "gap" in his preliminary picture? Here the reader is referred to the text itself, i.e. to part III (p. 199). Specifically, the reader learns of the following mental operations (ideas):

> Of course operations of traditional logic are involved, such as induction, inference, formulating theorems, derivation—and observation and ingenious experimentation. (...) But all these operations function in their place in the *whole* (gestalt-like) process. The process itself is governed by the *recentering* which originates in the desire to get comprehensive insight. (...) Fundamental changes in concepts were involved in the transition from the old view to the new. (...) For example, previously, test and certain "natural" circular movements were fundamentally separated from, opposed to, the other movements. Now rest and constant rectilinear movement are seen together ... in the symmetrical division of the whole picture.—(P. 201; emphasis: V.S.)

Notably the meaning of the various motion concepts changed during Galileo's thought processing "from the old view to the new;" indeed: "The meanings of the concepts about motion were radically changed in their place, whole, and (gestalt) function. (...) The new concepts opened up a most marvelous vista for understanding a large number of phenomena (in motion physics). Newton subsequently could view these movements as generated by inertia, on the one hand, tending toward rectangular movement, and by gravitation, on the other" (pp. 201f). At this point Wertheimer adds his general evaluation of such "*delightful* experiences" like that of Galileo's genious discovery: "Two directions are involved: getting a whole consistent picture, and seeing what the structure of the whole requires for the parts (of a thought process)" (p. 202).

Einstein: The Thinking That Led to the Theory of Relativity (Chap. VII)

> Those were wonderful days, beginning in 1916, when for hours and hours I was fortunate enough to sit with Einstein, alone in his study, and hear from him the story of the dramatic developments which culminated in the theory of relativity. During those long discussions I questioned Einstein in great detail about the concrete events in his thought.—(P. 203)

Wertheimer's famous chapter on Einstein's theory of relativity is divided into three parts (I-III), and ten acts for part I (pp. 203–233). It deals mainly with the crucial ideas of the theory of special relativity by Albert Einstein (1879–1955).—Act I ("The beginning of the problem"): Wertheimer introduces act I as follows: "The (thinking) process started in a way that was not very clear, and is therefore difficult to describe. First came such questions as: What if one were to run after a ray of light as it travels, would its velocity thereby be decreased? If one were to run fast enough, would it no longer move at all? . . ." (p. 204). Einstein's desire to plan some respective experiments were couched with doubts: "His wish to design such experiments was always accompanied by some doubt that the thing was really so; in any case, he felt that he must try to decide (something) . . ." (p. 204).—Act II ("Light determines a state of absolute rest?"): Wertheimer asserts that the layman, not aquainted with modern physics ". . . would not be able to follow the topics (acts) II and III . . .; the lay reader may therefore pass on to Act IV" (p. 205); he then adds the following:

> Back of all this there had to be something that was not yet grasped, not yet understood. Uneasiness about this characterized (already) young Einstein's state of the mind. When I asked him whether, during this period, he had already had some idea of the constancy of light velocity, independent of the movement of the reference system, Einstein answered decidely: "No, it was just *curiosity*. That the velocity of light could differ depending upon the movement of the observer was somehow characterized by *doubt*. Later developments increased that *doubt*."—(P. 205; emphasis: V.S.)

Act IV ("Michelson's result and Einstein"): In this act, Einstein ponders about the physicists's confrontation with the result of Michelson's former work: "(Michelson's) experiment was repeated, and the negative result was clearly confirmed. In fact the result contradicted all the reasonable expectations" (p. 207). But he went on and asked himself: "Precisely how does this (contradictory) result come about? The problem was an *obsession* with Einstein although he saw no way to a positive solution" (p. 208; emphasis: V.S.).—Act VI ("Re-examination of the theoretical situation"): As Einstein said to himself: "Except of that (negative) result, the whole situation in the Michelson experiment seems absolutely clear; all the factors involved and their interplay seem clear. But *are* they really clear?" And Wertheimer remarks to this latter question: "During this phase he was often depressed, sometimes in despair . . ." (p. 208).—Act VII ("Positive steps toward clarification"): Einstein's thoughts regarding the issues of "observability" and exact "measurement" of time-space relationships are treated; for instance: "For what now followed in Einstein's thinking we can fortunately report paragraphs from his own writing. He wrote them in the form of a discussion with the reader" (p. 209, cf. also fn 2). Accordingly, Einstein reached at one of his now classical conclusion: "For the observer within the (given) movement system will get results which differ from those of an observer in another frame of reference (namely): Every system has its special time and space values. A time or space judgment has sense only if we know the system of reference to which the judgment was made" (p. 212).

Act VIII ("Invariants and transformation"): In this act the concepts of "frame of reference," "time measurement" as well as physically relevant "constants" played a major role—for instance, as Einstein asked himself: "Can a relation between space and time of events in systems which move linearly to each other be so conceived that the velocity of light becomes constant? Eventually Einstein reached the answer: Yes!" (p. 213).—Wertheimer continued asking Einstein:

> In the discussions I had with Einstein in 1916 I put this question to him: "How did you come to choose just the velocity of light as a constant? Was this not arbitrary ... simply to fit in with these (earlier physicists´) experiments and with the Lorentz transformation?" Einstein's first reply was that we are entirely free in choosing axioms. (...) But then Einstein himself smilingly proceeded to give me a very nice example of an unreasonable axiom (...).— (P. 214; cf. also pp. 217ff)

Act IX ("On movement, on space, a thought experiment"): A comparison is made between Isaac Newton's theory of motion and Einstein's own theoretical notion; thereby the issue of "absolute" versus "relative" movement in physics is treated thus finding the basis for an *experimental* approach in the present context of inquiry (pp. 215f).—Act X ("Questions for observation and experiment"): Wertheimer kept asking Einstein if he could suggest further *decisive* experiments; for example: "Is it possible to find crucial physical questions to be answered in experiment that will decide whether these new theses are ´true´; whether they fit facts better, give better predictions of physical events than the old theses? He found a number of such crucial experiments, some of which physicists could and later did carry out" (p. 216). In the end, Einstein's theoretical issues led him to his further statement, i.e. "... to the question of the general theory of relativity" (parts II & III, cf. pp. 216–218). Wertheimer notes that further asking became mandatory for him: "The psychologist ... is concerned with the problems of thinking, what went on psychologically: ... Einstein said once in this context, whether there can be a way of really understanding the mircacle of thinking: ´Certainly you are right in trying to get at a deeper understanding of what really goes on in a thinking process ...´" (p. 217, cf. fn 5). And Wertheimer went on asking: "Why just these (mental) operations? Why was Einstein moving in just this direction. How did it come about that after he made one step, he followed just with that other step? (...) How did the new axioms arise? Did Einstein just try any axioms of which certain ones the actually happened to work? ... (In fact) the axioms were only a matter of later fomulations—after the real thing, the main discovery, had happened" (pp. 217ff; cf. also fn 7). In the end of the chapter he asserts that Einstein's thinking remained clear in the total developmental processing of his theorizing; and he adds the following: "In later years Einstein often told me about the problems on which he was working at the time. There was never a blind step. When he dropped any direction, it was only because he realized that it would introduce ununderstandable (i.e. incomprehensible: V.S.) arbitrary factors. . . ." (p. 222).

In his last letter to Albert Einstein dated August 9, 1943 Wertheimer was still not sure concerning some passages in his chapter, and he requested its clarification. Einstein's response to him on September 9, 1943 was positive: he agreed with Wertheimer's treatment of his theory (Sarris & Michael Wertheimer, 2018; cf. Max Wertheimer Archives at the University of Colorado in Boulder, CO). The book ends with the presentation of the major *gestalt* messages ("Conclusion: dynamics and logic of productive thinking"; pp. 224–250)—it is a summary which consists of Wertheimer's Credo of the perspectives of productive thinking and, in addition, of an abstract of the main notions in the book. Thereby his key concept of "transposition" is treated by means of a musical-notes illustration (pp. 242ff; cf. Fig. 150, A through D; see p. 243).

Wertheimer's Credo

(1) In contrast to ordinary thought processes in daily life, gestalt theory suggests that there are genuine, clean, and productive processes which govern human cognition. It is not true that men do not enjoy "productive" reasoning—on the contrary, "this (idea) is a result that may be deeply appreciated" (p. 224).

(2) The productive processes rely on factors which have not been considered, and accounted for, as being relevant in earlier classical models or theories of cognition. In fact, the nature of the productive operations in thinking—e.g., grouping and reorganization (cf. Table III, pp. 225f)—is alien to the traditional approaches.

(3) The truely "productive" features are not piecemeal ("blind") because they are related to the socalled whole-characteristics: ". . . the items, data, relations, etc., arise and function as parts in their place and role within the whole" (pp. 224f: see also Wertheimer's classical motto as cited in the beginning of the present *Introduction*; see p. 1).

(4) Although some obstructive factors and operations, in the sense of the earlier traditional interpretations, are or may be partly involved in a given thought process, nevertheless they "likewise function in relation to the whole-characteristics;" in this sense, they are context-dependent according to the general principle of "transposition" (pp. 242–244).

(5) The productive processes, which are *not* of the character of the socalled "and-summative" and "piecemeal" nature, are not arbitrary in nature; on the contrary: ". . . in spite of difficulties, in spite of some deviations . . ., thought processes show a consistency of (dynamic) development" (p. 225).

(6) Truly productive thought processes lead to meaningful, i.e. sensible solutions, expectations, and future-oriented expectations and "they call for an honest attitude for verification toward truth; . . . otherwise, there is the danger of dilettantism, of cheap

plausibility (in thinking)" (p. 225; see Wertheimer's explicit reference to his article "*On truth*", 1934).

The following quotation may be of help in appreciating this Credo's most relevant message: "Men are unhappy if the complication of (the piecemeal & and-sum) features befogs (i.e. obscures: V.S.) the issue (of productive thinking); they long for a structurally clear view in which the (single) items find their clear place, function and role, do not disturb the main lines and the resulting direction of view and action" (p. 235; see below Fig. 2).

Reception of the Book

Wertheimer's friend Albert Einstein allegedly suggested that a good scientist daily ponders for at least half an hour about the potential contra-positions to his or her professional collegues: How much may the recommendation hold also for Wertheimer's work?

After its publication (1945) the book "Productive Thinking" has been reviewed in some fifteen scientific, educational, philosophical, and literary magazines in Northamerica: from the *Psychological Bulletin, The American Journal of Psychology, Philosophical Review*, and the *New York Times*. With a few exceptions the critical reactions were largely positive; however, a few times the book's "vagueness" regarding gestalt theory was noted. Furthermore, Brett King and Michael Wertheimer (2005, p. 351) have pointed to one particular commentary which was critical, namely: Max Wertheimer "assumes that any of his (many) examples can without undue distortion be taken as instances of *one basic process*" (emphasis: V.S.; see below, "Wertheimer's book from a modern perspective").

A sign of the book's overall positive reception is the fact that it became republished several times in the USA (second enlarged edition, 1959, third enlarged edition, 1982; cf. Ericsson, Polson & Michael Wertheimer, 1982); and it was translated into the Japanese (1952) and also into the German (1957, 1964) language. Due to the wellknown *Zeitgeist* changes in the perspectives of a work, Wertheimer's book should be evaluated in line with the developments of psychology during the last 75 years, for instance: What is and remains the "revolutionary" impact of this work since the last decades of psychology (e.g., Murray, 1995, Kaufman & Sternberg, 2019)? Also, how could the more applied issues of today and tomorrow be evaluated in a "productive" sense of meaning ("*when the scales fall from your eyes*", so to speak; cf. Preface, p. vii)?

Wertheimer's Book from a Modern Perspective

Besides the fact that mathematics in our modern school systems may be still one of the least liked subjects, Wertheimer's psychological analyses of "productive" thinking in geometrics and algebra are all the more of great interest—especially when one points to the various forms of *contra*-productive, i.e. stultifying computer games and growing tendencies of

Fig. 2 Cover sheet of Wertheimer's book "*Productive Thinking*", with the photo of Max Wertheimer and his wife Anna together with their children Valentin, Lise, and Michael (from left to right); the photo was taken during their arrival in New York on September 16, 1933. See main text, above. Source: Sarris, 1996

internet addiction of the youth. However, the book does not present a systematic treatment of gestalt theory: In Wertheimer's preface it is stated that it offers merely a sort of first introduction ("prolegomena"): two further books should follow, i.e. one on the socalled "gestalt logic", the other on some systematic aspects of psychology—but the two books never appeared (see p. 33, cf. also King & Michael Wertheimer, 2005, p. 354).

Nevertheless the book provides a true challenge also for today's psychology of thinking: The cognitive psychologists of our time who work in such fields of creative and goal-oriented thinking are still faced with the complexities of the dynamics of human cognition and its relations to the socalled "fast" versus "slow" types of thinking (Kahneman, 2011); and they are confronted more than ever with the multitude of unresearched issues in learning and memory (short- and longterm memory; cf. the unanswered linkages concerning motivation and emotions in creative thinking: e.g., see Kaufman & Sternberg, 2019). Or as another example of the cognitive complexities involved, face this question: How is the thesis of the wellknown perceptual and cognitive distortions substantially related to Wertheimer's "productive" thinking processing? Perhaps Irvin Rock's assumption that perceptual and thinking processes are deeply intertwined is still valid, namely: "… it is entirely possible that we may learn about the operations of thinking by studying perception" (Rock, 1983). It is also important to ask if, and to which extent, the perceptual and thinking issues are connected with the developmental as well as comparative processes in humans and other animals (Sarris, 2006).

From our today's point of view Wertheimer's perspectives in the treatment of productive thinking may often seen as enthusiastic and idealistic; however, consider the **caveat** in the end of his book (productive versus *contra*-productive factors), which must not be overlooked: "Of course there are often strong external factors working against these (productive) processes as, e.g., blind habits, certain kinds of school drill, (personal) bias, or special interests" (p. 189). At any rate, in the last few decades there has been a host of follow-up research, especially in the fields of cognitive heuristics and bounded rationality (Simon, 1978; Simon et al., 2008; Gogerenzer & Gaissmaier, 2011) and also in neuroscience and artificial intelligence (e.g., Kounios & Beeman, 2014).

Some further critical questions might be raised:

What about the quantitative-methodological relevance of Wertheimer's book from a today's perspective? For instance, to which extent do the productive thought processes—as assumed throughout in the work—really form a basic ***unity***; first of all, should there not be established a person-oriented typology of the different underlying (sub)processes? At any rate, in present-day research on creativity much more systematic and quantitative empirical work is done in comparison to earlier work relying mostly on a descriptive *post hoc* basis with individual case studies (N = 1). As a matter of fact, nowadays many researchers try and follow the much more rigorous demands of statistical-quantitative methodology, with the help of mathematical and computer modeling. And yet, Wertheimer's book remains relevant for tomorrow's psychologists and other scholars in neurobiology, education, and philosophy (Sternberg & Davidson, 1995; Knoblich & Öllinger, 2006)—like a durable *thorn in our flesh* for the future study of creativity (Kaufman & Sternberg, 2019). After all, this is important due to the fact that the potentials of human productive thinking appear to be still underdeveloped.

References

Arnheim, R. (1989). Max Wertheimer. *Psychological Research, 51*, 45–46.

Ash, M. G. (1989). Max Wertheimer's university career in Germany. *Psychological Research, 51*, 52–57.

Ash, M. G. (1995). *Gestalt psychology in German culture 1890–1967: holism and the quest for objectivity*. Cambridge, MA: Cambridge University Press.

Duncker, K. (1935). *Zur Psychologie des produktiven Denkens*. Berlin: Springer. (Translated into English, 1945).

Ehrenfels, C. von (1890). Über Gestaltqualitäten. *Vierteljahresschrift für wissenschaftliche Philosophie, 14*, 249–292.

Einstein, A. (1991). Vorwort von 1950 (Preface of 1950). In H.-J. Walter (Ed.), *Max Wertheimer: Zur Gestaltpsychologie menschlicher Werte* (pp. 9–10). Wiesbaden: Springer.

Ericsson, A., Polson, P. G. & Mich. Wertheimer (1982). Preface to the Phoenix edition (3rd ed.). In Mich. Wertheimer (Ed.), *Max Wertheimer: Productive Thinking* (pp. xi-xvii). New York: Harper.

Gigerenzer, G. & Gaissmaier, W. (2011). Heuristic decision making. *Annual Review of Psychology, 62*, 451–482.

Gundlach, H. (2014). Max Wertheimer, *habilitation* candidate at the Frankfurt Psychological Institute. *History of Psychology, 17*, 134–148.

Kahneman, D. (2011). *Thinking: fast and slow*. New York: Farrar.

Kaufman, C. & Sternberg, R.J. (Eds.). (2019). *Cambridge handbook of creativity* (2nd ed.). Cambridge, MA: Cambridge University Press.

King, D. B. & Mich. Wertheimer (2005). *Max Wertheimer & Gestalt theory*. New York: Transaction Publ. (New York: Routledge, 2017).

Kounios, J. & Beeman (2014). The cognitive neuroscience of insight. *Annual Review of Psychology, 65*, 71–93.

Knoblich, G. & Öllinger, M. (2006). Einsicht und Umstrukturierung beim Problemlösen. In J. Funke (Ed.), *Enzyklopädie der Psychologie. Denken und Problemlösen* (Bd. 8, S. 1–85). Göttingen: Hogrefe.

Murray, D. J. (1995). Gestalt ideas about thinking and problem-solving. In D. J. Murray, *Gestalt psychology and the cognitive revolution* (pp. 132–164). New York: Harvester.

Mussweiler, T. (2003). Comparative processes in social judgment: Mechanisms and consequences. *Psychological Review, 110*, 472–489.

Newman, E. B. (1989). Remembering Max Wertheimer 1931-1943. *Psychological Research, 51*, 47–51.

Rock, I. (1983). *The logic of perception*. Cambridge, MA: MIT Press (Cited in Sarris, 1995).

Sarris, V. (1987). Max Wertheimers Frankfurter Arbeiten zum Bewegungssehen – die experimentelle Begründung der Gestaltpsychologie. *Forschung Frankfurt, 5*, 17-23. (Reprinted in W. Meißner, Ed., *Forschung Frankfurt: Sonderband zur Geschichte der Universität*; pp. 120-126. Frankfurt a. M.: Goethe-Universität, 2000.)

Sarris, V. (1989). Max Wertheimer on seen motion: theory and evidence. *Psychological Research, 51*, 58–68.

Sarris, V. (1995). *Max Wertheimer in Frankfurt: Beginn und Aufbaukrise der Gestaltpsychologie*. Lengerich (Germany): Pabst.

Sarris, V. (1996). Max Wertheimer in New York: Zum 50. Jahrestag von *Productive Thinking (1945). Psychologische Rundschau, 47*, 137–145.

Sarris, V. (1997). Reflexionen über den Gestaltpsychologen Max Wertheimer und sein Werk: Vergessenes und wieder Erinnertes. In M. Hassler & J. Wertheimer (Eds.), *Der Exodus aus Nazi-Deutschland und die Folgen: Jüdische Wissenschaftler im Exil* (pp. 177–190). Tübingen: Attempto.

Sarris, V. (2006). *Relational psychophysics in humans and animals: a comparative-developmental approach.* London: Psychology Press.

Sarris, V. (2012). Epilogue: Max Wertheimer in Frankfurt and thereafter. In L. Spillmann (Ed.), *Max Wertheimer on perceived motion and figural organization* (pp. 253–265). Cambridge, MA: MIT Press.

Sarris, V. & Mich. Wertheimer. (1987). Max Wertheimer (1880-1943) im Bilddokument – ein historiografischer Beitrag. *Psychologische Beiträge, 29*, 469–493.

Sarris, V. & Mich. Wertheimer. (2001). Max Wertheimer's research on aphasia and brain disorders: a brief account. *Gestalt Theory, 23*, 267–277.

Sarris, V. & Mich. Wertheimer. (2018). Max Wertheimer: Productive Thinking (1945). In H. E. Lück, R. Miller & G. Sewz (Eds.), *Klassiker der Psychologie* (2nd ed., pp. 158–162). Stuttgart: Kohlhammer.

Sekuler, R. (1996). Motion perception: a modern view of Wertheimer's 1912 monograph. *Perception, 25*, 1243–1258. (Reprinted in L. Spillmann, Ed., *Max Wertheimer on perceived motion and figural organization*, pp. 101–125; Cambridge, MA: MIT Press, 2012).

Simon, H. A. (1978). Information-processing theory of human problem solving. In W. K. Estes (Eds.), *Human information processing* (pp. 271–295). Hillsdale, NJ: Erlbaum.

Simon, H. A., Egidi, M., Viale, R. & Marris, R. (Eds.). (2008). *Economics, bounded rationality and the cognitive revolution*. London: Edward Elgard Publ. (Internet access: 27/04/2019).

Spillmann, L. (Ed). (2012). *Max Wertheimer on perceived motion and figural organization.* Cambridge, MA: MIT Press.

Sternberg, R. J. & Davidson, J. E. (Eds.). (1995). *The nature of insight*. Cambridge, MA.: MIT Press.

Stock, A. (2014). Schumann's wheel tachistoscope: its reconstruction and its operation. *History of Psychology, 17*, 149–158.

Wertheimer, Mich. (2014). Music, thinking, perceived motion: the emergence of gestalt theory. *History of Psychology, 17*, 131–133.

Westheimer, G. (1999). Gestalt theory reconfigured: Wertheimer's anticipation of recent developments in visual neuroscience. *Perception, 28*, 5–16.

Publications of Max Wertheimer (an updated list of Wertheimer's original publications is contained elsewhere; cf. Spillmann, 2012)

Wertheimer, M. (1905). Experimentelle Untersuchungen zur Tatbestandsdiagnostik. *Archiv für die gesamte Psychologie, 6*, 59–131. (Also published separately, Leipzig: Engelmann, 1905; pp. 1–70).

Wertheimer, M. (1910). Musik der Wedda. *Sammelbände der internationalen Musikgesellschaft, 11*, 300–309.

Wertheimer, M. (1912a). Über das Denken der Naturvölker: 1. Zahlen und Zahlgebilde. *Zeitschrift für Psychologie, 60*, 321–378.

Wertheimer, M. (1912b). Experimentelle Studien über das Sehen von Bewegung. *Zeitschrift für Psychologie, 61*, 161–265. (Translated into English: *Max Wertheimer on perceived motion and figural organization*, pp. 1–91; L. Spillmann, ed., Cambridge, MA: MIT Press, 2012).

Wertheimer, M. (1920). *Über Schlußprozesse im produktiven Denken*. Berlin: de Gruyter. (Abridged translation in W.D. Ellis, ed., *A sourcebook of Gestalt psychology*, pp. 274–282; New York: Harcourt, Brace, 1938).

Wertheimer, M. (1923). Untersuchungen zur Lehre von der Gestalt. II. *Psychologische Forschung, 3*, 106–123. (Translated into English: *Max Wertheimer on perceived motion and figural organization,* pp. 127–182; L. Spillmann, ed., Cambridge, MA: MIT Press, 2012).

Wertheimer, M. (1925). Über Gestalttheorie. (Vorlesung in der Kant Gesellschaft, 17. Dezember 1924). *Philosophische Zeitschrift für Forschung und Aussprache, 1*, 39–60. (Translated into English: *Social Research*, 1944, *11*, 78–99).

Wertheimer, M. (1934). On truth. *Social Research, 1*, 135–146. (Reprinted in M. Henle, Ed., *Documents of Gestalt psychology*. Los Angeles: University of California Press, 1961).

Wertheimer, M. (1935). Some problems in the theory of ethics. *Social Research, 2*, 353–367. (Reprinted in M. Henle, Ed., *Documents of Gestalt psychology*. Los Angeles: University of California Press, 1961).

Wertheimer, M. (1937). On the concept of democracy. In M. Ascoli & F. Lehman (Eds.), *Political and economic democracy* (pp. 271–283). New York: Norton. (Reprinted in M. Henle, Ed., *Documents of Gestalt psychology*. Los Angeles: University of California Press, 1961).

Wertheimer, M. (1940). A story of three days. In R. N. Anshen (Ed.), *Freedom: its meaning* (pp. 555–569). New York: Harcourt, Brace. (Reprinted in M. Henle, Ed., *Documents of Gestalt psychology*. Los Angeles: University of California Press).

Wertheimer, M. (1945). *Productive thinking*. New York: Harper. (Translated into Japanese by T. Yatabe, and published in Tokyo by I.G. Sosho, Tokyo 1952; translated into Germany by W. Metzger, and published in Frankfurt a. M. by W. Kramer, 1957; 2nd ed., 1964. – Enlarged English edition by Mich. Wertheimer, New York: Harper, 1959; reprinted & enlarged 1982, University of Chicago Press).

Productive Thinking (1945)

2

Max Wertheimer

M. Wertheimer (✉)
New York, NY, USA

M. Wertheimer, *Max Wertheimer Productive Thinking*, Classic Texts in the Sciences,
https://doi.org/10.1007/978-3-030-36063-4_2

PRODUCTIVE THINKING

MAX WERTHEIMER

LATE PROFESSOR OF PSYCHOLOGY AND PHILOSOPHY IN THE GRADUATE FACULTY OF POLITICAL AND SOCIAL SCIENCE, THE NEW SCHOOL FOR SOCIAL RESEARCH

HARPER & BROTHERS PUBLISHERS

NEW YORK AND LONDON

K-W

CONTENTS

CONTENTS

CONTENTS

ACKNOWLEDGMENTS

This book owes much to Clara W. Mayer, Dean of the School of Philosophy and Liberal Arts in the New School for Social Research. But for her untiring effort the manuscript would not have reached its final form; she was intensely interested in its subject matter and deeply devoted to its aims. That she never failed to find time for it notwithstanding her full time activities at the School, was a source of inspiration.

I am very grateful to Dr. S. E. Asch for his fine help in preparing the manuscript; to Mr. and Mrs. Benno Elkan for untiring friendly encouragement; to Mrs. Clara Mond and Mrs. Maria Di Piazza for indefatigable secretarial assistance.

Special thanks are due Dr. Alvin Johnson and my Faculty. I regret that this book, which is limited to some elementary problems, does not fully express the spirit of co-operation in the social sciences which is so alive in our Faculty and which owes so much to Alvin Johnson. Only briefly and occasionally could I touch upon these issues, which mean so much to all of us. I am likewise indebted to my friends whose work is connected with the problems of this book.

To the subjects of my experiments, the adults and the children, I am thankful for having learned so much. And I am very grateful to the distinguished men of science—Einstein above all—who made it possible for me to study intimately, in many conversations, how some of their great achievements in thinking developed.

The book contains only a small portion of the studies and material I have given to students in lectures and seminars. I have reported on parts of it in single lectures at Harvard, Chicago,

ACKNOWLEDGMENTS

Bloomington, Ann Arbor, Smith College, Swarthmore, Bryn Mawr, et al.

It is a kind of prolegomena; I hope I shall be able to finish two other books to which this is a mere introduction. In spite of its limitations I hope it may be of use.

MAX WERTHEIMER

New Rochelle, N. Y.
September 23, 1943

NOTE BY THE EDITORS

THE manuscript of *Productive Thinking*, upon which Max Wertheimer had worked during the last seven years, was completed shortly before his death. In preparing the work for publication we found that certain linguistic revisions were necessary. Mere difficulties of expression, it seemed to us, should not be allowed to obscure the meaning which Wertheimer clearly had in mind. The reader may be assured, however, that changes have been made only where the purpose of clarity could not otherwise have been attained. The content and the form of the discussion have not been altered in any place.

All remaining notes and papers of Wertheimer will be deposited with the Graduate Faculty of Political and Social Science, where qualified scholars will have the opportunity to study them.

Valentin Wertheimer assisted his father in some of the final revisions of the manuscript and has been helpful to us in many ways.

We are grateful to Mrs. Ray Borne for the painstaking care with which she saw the manuscript through the press.

S. E. Asch
W. Köhler
C. W. Mayer

PRODUCTIVE THINKING

INTRODUCTION

WHAT occurs when, now and then, thinking really works productively? What happens when, now and then, thinking forges ahead? What is really going on in such a process?

If we look for answers in books, we often find apparently easy ones. But confronted by actual processes of this kind—when one has just had a creative idea, however modest the issue, when one has begun really to grasp an issue, when one has enjoyed a clean, productive process of thought—those answers often seem to cover up the real problems rather than to face them squarely. The flesh and blood of what has happened seem to be lacking in those answers.

Surely in the course of your life you have been curious about a lot of things, sometimes seriously. Have you been equally serious about what this thing called thinking may be? There are, in this world of ours, eating, thunderstorms, blossoms, crystals. Various sciences deal with them; they attempt by great effort to get real understanding, to grasp what these things really are. Are we equally serious when we ask what productive thinking is?

There are fine cases. You can find them often, even in daily life. If you have had your eyes open, you have probably encountered somewhere in your life—if nowhere else, then in children—this surprising event, the birth of a genuine idea, of a productive development, the transition from a blind attitude to understanding in a productive process. If you have not been fortunate enough to experience it yourself, you may have encountered it in others; or you may—fascinated—have glimpsed it when reading good books.

Many are of the opinion that men do not like to think; that they will do much to avoid it; that they prefer to repeat instead. But in spite of many factors that are inimical to real thinking, that suffocate it, here and there it emerges and flourishes. And often one gets the strong impression that men, even children, long for it.

What really takes place in such processes? What happens if one really thinks, and thinks productively? What may be the decisive features and the steps? How do they come about? Whence the flash, the spark? What are the conditions, the attitudes, favorable or unfavorable to such remarkable events? What is the real difference between good and bad thinking? And in connection with all these questions: how improve thinking? your thinking? thinking itself? Suppose we were to make an inventory of basic operations in thinking—how would it look? What, basically, is at hand? Could the basic operations themselves be enlarged and improved, and thus be made more productive?

For more than two thousand years some of the best brains in philosophy, in logic, in psychology, in education, have worked hard to find real answers to these questions. The history of these efforts, the brilliant ideas brought forward, the hard work done in research and in theoretical discussion, present on the whole a rich, dramatic picture. Much has been achieved. In a large number of special questions solid contributions to understanding have been made. At the same time there is something tragic in the history of these efforts. Again and again when great thinkers compared the ready answers with actual, fine thinking, they were troubled and deeply dissatisfied—they felt that what had been done had merits, but that in fact it had perhaps not touched the core of the problem at all.

The situation is still somewhat of this kind. To be sure, many books deal with these questions as if, fundamentally, everything were settled—in one way or another. For there are basically different ideas about what thinking is, each with serious consequences for behavior, for education. When observing a teacher we may often realize how serious the consequences of such ideas about thinking can be.

Although there are good teachers, with a natural feeling for what genuine thinking means, the situation in schools is often not good.

How teachers act, how a subject matter is taught, how textbooks are written, all this is widely determined by two traditional views about the nature of thinking: the view of traditional logic and the view of association theory. These two views have their merits. To a degree they seem adequate to certain types of thought processes, to certain jobs in thinking; but it is at least an open question whether the way in which they interpret thinking does not cause serious hindrance, an actual impairment of genuine abilities.

This book has been written because the traditional views have ignored important characteristics of thought processes, because in many other books those views are taken for granted without real investigation, because in such books the discussion of thinking runs largely in mere generalities, and because, for the most part, the gestalt view is only superficially known. Much is at stake and it seems proper to bring these neglected issues to the fore, to examine the traditional views, to discuss the crucial problems in concrete instances of fine, productive thought, and in doing so, to give the gestalt interpretation of thinking.

In a number of chapters (I-III) elementary, seemingly obvious examples will be used for the discussion. We shall face the basic theoretical issues in direct contact with the concrete material. Several methods of experimental discussion will contribute to clarification. We shall study how thinking actually proceeds and what the nature of the process is as a whole, as well as in its parts, steps, and operations. In contrast to poorer ways of thinking the reader may enjoy the way in which fine, though modest, productive processes occur in children.

We shall see that what is going on in these processes is far from being adequately handled by the tools and views of the two traditional approaches. We shall recognize characteristics and operations that have been ignored because they are intrinsically foreign to the customary conceptions. We shall see how such factors work in the achievement of thinking.

In Chapter IV we shall describe a modest example from human life, which seems to touch the essence of profound human issues.

We shall also give in Chapters III, V, VI, and VII several descriptions and interpretations of greater thought processes, concluding with the story of the thought process that led Einstein to the discovery of the relativity theory. In the final chapter we shall formulate our general conclusions.

The expert knows how many conditions must be met in a careful investigation. I shall have to omit the report of many technical points that are important in research, because they would make the report too cumbersome. In an investigation one often encounters things one is at first disposed to understand in some traditional way. Closer scrutiny then shows that the matter is not so easily settled. As a consequence one looks for ways, for methods that may yield deeper clarification. The scientific reader would be interested in such specific methods and techniques, the nature of the steps taken in the theoretical and experimental discussion. But the first need is for careful observation and investigation in qualitative experimentation. To be sure, in many cases, it was easy to change the qualitative into a quantitative method, which for many problems is needed as a second step; however, I shall not report on them in this book.

To the scientific psychologist, to the logician, to the educator this book is offered primarily as an invitation to reconsider basic issues. I have chosen a terminology which seemed closest to the nature of the actual processes. Although much of what I am going to say comes, I think, very near genuine common sense, it is difficult to express in scientific terms, and the terms which I use may often seem strange to the reader because they are at odds with his customary ways of viewing the problems. My terms should not give the impression that the problems are settled; they are themselves loaded with—I think—productive problems. The terms and the theses are rather to be understood as arrows, pointing first of all to the concrete steps and characteristics that actually occur in the examples. Much of what I shall say could be expressed in other terminology. Many of the problems and theses are to a degree neutral to expressions of one sort or another. The terminology itself does not matter. What does matter are the problems and the nature of the theses as they vividly appear in the discussion of the concrete cases. Currently some of the terms and theses are being used in a superficial way—the concreteness of the discussion may help to clear up misunderstandings which might otherwise follow.

Notwithstanding the possibility of expressing the facts in other language, including the language of other approaches, let me caution the scientific reader: the trend indicated by this research leads to a view that is basically in contrast with many current views. I hope the reader will not rest content with storing that view in some drawer in which he classifies psychological or philosophical opinions. More is at stake. We have to face these issues in a concrete and productive fashion.

As a kind of background for the following discussions, I present first a very short characterization of the two traditional approaches. They surpass all others in the rigor and completeness with which they consider operations and establish basic concepts, standards, criteria, laws and rules. Other approaches—even if they seem at first in strong opposition to these two—often still contain as their very meat, in one way or another, precisely the operations, the rules of these two. Modern research in thinking is largely determined by one or the other, or both at the same time. I shall indicate their main lines, but shall omit some points which appear as additions of another nature and which, besides, are not clear in themselves.

I. Traditional logic attacked the problems in an ingenious fashion: how are we to find the main issues in the vast variety of the topics of thinking? As follows: thinking is concerned with truth. Being true or false is a quality of assertions, propositions, and only of these. The elementary form of proposition asserts or denies some predicate of a subject, in the form "all S are P," or "no S is P," or "some are," or "some are not." Propositions involve general concepts—class concepts. These are basic to all thinking. For the correctness of a proposition it is decisive that its "intension" or "extension" be dealt with correctly. On the basis of assertions inferences are drawn. Logic studies formal conditions under which inferences are or are not correct. Certain combinations of propositions make it possible to derive "new," correct propositions. Such syllogisms, with their premises and their conclusions, are the crown, the very heart of traditional logic. Logic establishes the various forms of syllogism which guarantee correctness of the conclusion.

Although most of the textbook syllogisms seem barren, a kind of circle, like the classical example—

All men are mortal
Socrates is a man
therefore, Socrates is mortal—

there are examples of real discoveries which can in a first approach be regarded as syllogisms, as for example the discovery of the planet Neptune. But formally, basically, there seems to be no real difference between the two kinds of syllogism.[1] The decisive characteristics and the rules are identical for both—the somewhat silly and the really sensible ones.

Traditional logic is concerned with the criteria that guarantee exactness, validity, consistency of general concepts, propositions, inferences and syllogisms. The main chapters of classical logic refer to these topics. To be sure, sometimes the rules of traditional logic remind one of an efficient police manual for regulating traffic.

If we disregard differences of terminology, controversies of a subtle nature, we may list as characteristic the following operations of traditional logic:

TABLE I

definition
comparison and discrimination
analysis
abstraction
generalization
forming class concepts
subsumption, etc.
forming propositions
forming inferences
forming syllogisms, etc.[2]

These operations as conceived, defined, and utilized by the logician have been and are being taken by psychologists as subjects

[1] See M. Wertheimer, "Über Schlussprozesse im produktiven Denken," in *Drei Abhandlungen zur Gestalttheorie* (Erlangen, 1925) pp. 164-184; see also W. D. Ellis, *A Source Book of Gestalt Psychology*, selection 23 (Harcourt, Brace & Company, 1939).

[2] The nature of these operations has been minutely discussed. For our purposes it does not matter much whether they are defined in mentalistic, behavioristic, pragmatic, or other language, although philosophically there are great differences among these various views.

for investigation. As a result, we have many experimental investigations on abstraction, generalization, definition, drawing conclusions, etc.

Some psychologists would hold that a person is able to think, is intelligent, when he can carry out the operations of traditional logic correctly and easily. The inability to form general concepts, to abstract, to draw conclusions in syllogisms of certain formal types is viewed as a mental deficiency, which is determined and measured in experiments.[3]

However one may view classical logic, it had and has great merits:

- in the decisiveness of its will to truth;
- in the concentration on the basic difference between a mere assertion, a belief and an exact judgment;
- in its emphasis on the difference between hazy concepts, hazy generalizations, and exact formulations;
- in the development of a host of formal criteria which are suited to testing for, and discovering mistakes, haziness in thinking such as unjustified generalization, jumping at conclusions;
- in its emphasis on proof;
- in the seriousness of the rules of discussion;
- in the insistence on stringency and rigor in each individual step in thinking.

The system of traditional logic, as envisaged in its main lines in the *Organon* of Aristotle, was recognized as final through the centuries; elaborations were added here and there, but these did not change its main character. A new branch started at the time of the Renaissance, a development that was essential to the growth of modern science. The central point was the introduction, as fundamental, of a procedure which until then had been regarded as of minor value because of lack of complete conclusiveness. This is the procedure of induction, with its emphasis on experience and ex-

[3] Some students nowadays think that traditional logic has no connection with actual *behavior*. This is a mistake. For application to behavior merely presupposes a connecting axiom, approximately as follows: behavior will be unreasonable, will fail of achievement, will run into trouble, if it is determined by factors parallel to mistakes in the sense of traditional logic.

perimentation, a methodological concept which reached its greatest perfection in John Stuart Mill's famous canon of rules of induction.

Ia. The emphasis here is not on rational deduction from general propositions but on gathering facts, on studying the empirically constant connections of facts, of changes, and on observing the consequences of changes introduced into factual situations, procedures which culminate in general assumptions.[4] Syllogisms are viewed as tools by which one can draw consequences from such hypothetical assumptions in order to test them.

It is widely believed that inductive logic adds to the classical rules and operations the emphasis on:

TABLE I*a*

empirical observations
careful gathering of facts
studying problems empirically
introducing experimental methods
correlating facts
developing crucial tests

II. The second great theory of thinking is centered in the classical theory of associationism. Thinking is a chain of ideas (or, in more modern terms, a chain of stimuli and responses, or a chain of behavior elements). The way to understand thinking is clear: we have to study the laws governing the succession of ideas (or, in modern terms, of behavioral items). An "idea" in classical association theory is some remnant of perception, a copy, in more modern terms, a trace of stimulations. What is the fundamental law of the succession, of the connection of these items? Answer—very elegant in its theoretical simplicity: if two items, *a* and *b*, have often occurred together, a subsequent occurrence of *a* will call forth *b* in the subject.[5] Basically the items are connected in the way in which my friend's telephone number is connected with his name, in which nonsense syllables become reproducible when learned in a series of such syllables, or in which a dog is conditioned to respond with salivation to a certain musical sound.

[4] The main point was the study of the correlation of two series of varying items, with resulting laws of function instead of mere classifications.

[5] In later scientific developments the law had to be qualified in certain details.

Habit, past experience, in the sense of items repeated in contiguity—inertia rather than reason, are the essential factors, just as David Hume had maintained. As compared with classical associationism, this theory is now being developed in a most intricate way; but the old idea of repetition, in contiguity, is still the central feature. A leading exponent of this approach stated explicitly not long ago that the modern theory of the conditioned reflex is essentially *of the same nature* as classical associationism.

The list of operations here looks about as follows:

TABLE II

association, acquiring connections—bonds on the basis of repetitions
role of frequency, of recency
recall from past experience
trial and error, with chance success
learning on the basis of repeated success
acting in line with conditioned responses, and with habit

These operations and processes are now being widely studied with highly developed methods.

Many psychologists would say: ability to think is the working of associative bonds; it can be measured by the number of associations a subject has acquired, by the ease and correctness with which he learns and recalls them.[6]

[6] Cf., e.g., E. L. Thorndike, *Psychology of Arithmetic* (Macmillan, 1922): "The pedagogy of the past made two notable errors in practice based on two errors about the psychology of reasoning. It considered reasoning as a somewhat magical power or essence which acted to counteract and overrule the ordinary laws of habit in man; and it separated too sharply the 'understanding of principles' by reasoning from the 'mechanical' work of computation . . . , remembering facts and the like, done by 'mere' habit and memory.

"Reasoning or selective, inferential thinking is not at all opposed to, or independent of, the laws of habit, but really is their necessary result under the conditions imposed by man's nature and training. A closer examination of selective thinking will show that no principles beyond the laws of readiness, exercise, and effect, are needed to explain it; that it is only an extreme case of what goes on in associative learning as described under the 'piecemeal' activity of situations. . . ." (p. 190).

Similarly, W. B. Pillsbury, in "Recent Naturalistic Theories of Reasoning" (*Scientia*, 1924): "The animal solves a problem by a series of chance trials . . . A scientific problem is solved in very much the same way by a number of chance thoughts" (p. 25). "One never can predict when a fruitful suggestion is to be made. Usually a number of inadequate suggestions will be made before

No doubt there are merits in this approach also, with regard to the subtle features at work in this kind of learning and behaving.

Both approaches had difficulties with regard to sensible, productive processes of thinking.

Consider first traditional logic. In the course of the centuries there arose again and again a deep-felt dissatisfaction with the manner in which traditional logic handles such processes.[7] In comparison with actual, sensible, and productive processes, the topics as well as the customary examples of traditional logic often look dull, insipid, lifeless. To be sure, the treatment is rigorous enough, yet often it seems barren, boring, empty, unproductive. If one tries to describe processes of genuine thinking in terms of formal traditional logic, the result is often unsatisfactory: one has, then, a series of correct operations, but the sense of the process and what was vital, forceful, creative in it seems somehow to have evaporated in the formulations. On the other hand it is possible to have a chain of logical operations, each perfectly correct in itself, which does not form a sensible train of thought. Indeed there are people with logical training who in certain situations produce series of correct operations which, viewed as a whole, nevertheless form something akin to a flight of ideas. Training in traditional logic is not to be disparaged: it leads to stringency and rigor in each step, it contributes to critical-mindedness; but it does not, in itself, seem to give rise to productive thinking.[8] In short, there is the danger of

---

the right one comes. They may be suggested by another person, even by a child or person ignorant of the problem. During the solution of the problem the thinker merely stands ready to pass upon the solutions offered. His attitude is very much like that which we may assume the animal to have as he works by trial and error. It is just as little controlled. It is, in fact, a trial and error process different from the other only in that the trials are made in imagination, not in real movements. . . . to find a way of obviating the difficulty. This is always by a series of trial and error processes, by a number of suggestions offered by association" (p. 30). It is only fair to add that in more recent publications Pillsbury sees the situation quite differently.

[7] Cf., for instance, certain movements against traditional logic at the end of the Middle Ages, or the marvelous fragment of young Spinoza, "Improvement of the Understanding." Tragic movements these were, prompted by a feeling of some basic inadequacy, but at the same time unable to achieve a really positive approach.

[8] The discussions of methodology in traditional logic, though meritorious in various respects, do not give real help at this point. Cf. the heuristic ideas (or also, veritable logical machines) of Buridanus, of Raimundus Lullus, of Jevons.

being empty and senseless, though exact; and there is always the difficulty with regard to real productiveness.

Realization of the latter point—among others—led in fact to the emphatic declaration by some logicians that logic, interested in correctness and validity, has nothing at all to do with factual thinking or with questions of productivity. A reason was also given for this: logic, it was said, has timeless implications and is, therefore, in principle, divorced from questions of actual thought processes which are merely factual and, of necessity, processes in time. This separation was certainly meritorious for certain problems; from a broader view, however, such assertions often look somehow like the declaration of the fox that the grapes were sour.

Similar difficulties arose in association theory: the fact that we have to distinguish between sensible thought and senseless combinations, and the difficulty in dealing with the *productive* side of thinking.[9]

If a problem is solved by recall, by mechanical repetition of what has been drilled, by sheer chance discovery in a succession of blind trials, one would hesitate to call such a process sensible thinking; and it seems doubtful whether the piling up of such factors only, even in large numbers, can lead to an adequate picture of sensible processes. In order to deal somehow with processes which reach new solutions, a number of auxiliary hypotheses were proposed (for instance, Selz's constellation theory, or the concept of the habit-family-hierarchy) which, by their very nature, do not seem to give decisive help.

In the past decades other views originated which brought new conceptions, new directions in the theory of thinking: e.g., the approach of Hegelian and Marxist dialectics which emphasize dynamic

[9] Characteristic in the former respect was the brilliant book of Hugo Liepmann on the flight of ideas (*Über Ideenflucht*, 1904). Discussing concrete examples of flight of ideas in mental patients, he found that the criteria offered by association theory do not really suffice to discriminate even between certain kinds of flight of ideas and reasonable talk.

A recent formulation exhibits the basic character of the modern form of association theory in a nutshell. I am referring to Clark Hull's paper on "Mind, Mechanism and Adaptive Behavior," *Psychological Review*, 1937, Vol. 44, pp. 1-32: "A *correct* or 'right' reaction is a behavior sequence which results in reenforcement. An *incorrect* or 'wrong' reaction is a behavior sequence which results in experimental extinction" (p. 15). One sees that the question

development in their doctrine of "inherent contradictions" with the three steps of thesis, antithesis, and synthesis; the broad development of logistics or mathematical logic (Whitehead, Russell *et al.*) which enriches the topics and operations of traditional logic by the study of the logic of relations, of relational networks, considers forms of conclusion other than syllogisms; phenomenology (Husserl) which stresses the viewing of essentials in "phenomenological reduction"; pragmatism (especially John Dewey) with its emphasis on doing and acting, instead of mere ghostlike thinking, on future and on actual progress; also in psychology—starting about the same time as the approach developed in this book—the "Denkpsychologie" of the Würzburg school (Kuelpe, Ach, Buehler, Selz, *et al.*), with the emphasis on "Aufgabe"—the role of a given task, on "thoughts" as "unanschauliche Vorstellungen," on relations, schemes etc.; the "naturalistic approach" (J. Dewey, Pillsbury, *et al.*) which centers on the conditions that start actual thinking in a given situation.

Most of these approaches are important in their philosophical and psychological aspects. Although in these developments the situation with regard to our main problem and the crucial points mentioned still seems far from satisfactory, some of them made really new contributions. Some again show the influence of the two classical approaches. In other words, if one penetrates through the new formulations to the nature of the operations which are actually posited *in concreto*, one finds to one's surprise that they are essentially operations of those two traditional approaches. This reminds one of cases that have frequently occurred in the history of logic. In the introduction or in some early chapter a book may seem to start a new approach, altogether different from the customary treatment of logic; in fact, certain formulations may almost appear akin to those of gestalt theory. And yet, when it comes to dealing concretely with a problem, the old operations, old rules, old attitudes appear again.

I have been able to do no more than briefly mention these approaches here. The expert will, I think, see in the text what in them

of repetition is *the* issue. These basic definitions are no doubt in line with the spirit of association theory.

is in line with and what is basically different in nature from the approach in this book.

This book focuses on some elementary, basic issues. The nature of the topics discussed permits us to deal with thought in terms of "relatively closed systems," as though thinking about a problem were a process that occurred independently of larger issues. Only occasionally shall we refer to the place, role, and function of such a process within the personality structure of the subject and within the structure of his social field. For the moment it will suffice if I remark that the same field principles discussed in this book also seem basic to an adequate treatment of processes within such larger regions.

CHAPTER I

The Area of the Parallelogram

AMONG the problems with which I worked was that of finding the area of the parallelogram.

I do not know whether you will enjoy as much as I did some of the experiences of which I am going to tell you here. I think you will, if you will travel with me on this exploratory journey on which problems arose and difficulties had to be faced for which I had to find tools and methods in order to clarify the psychological problems involved.

I

1. I am visiting a classroom. The teacher: "During the last lesson we learned how to find the area of a rectangle. Do you all know it?"

The class: "Yes." One pupil calls out: "The area of a rectangle is equal to the product of the two sides." The teacher approves, then gives a number of problems with rectangles of varying sizes, which all solve readily.

"Now," says the teacher, "we shall go on." He draws a parallelogram on the blackboard: "This is called a parallelogram. A par-

FIG. 1

allelogram is a plane quadrilateral the opposite sides of which are equal and parallel."

Here a pupil raises his hand: "Please, teacher, how long are the sides?" "Oh, the sides may be of very different lengths," says the teacher. "In our case one line measures 11 inches, the other 5 inches." "Then the area is 5 x 11 square inches." "No," answers the teacher, "That's wrong; you will now learn how to find the area of a parallelogram." He labels the corners *a, b, c, d.*

"I drop one perpendicular from the upper left corner and another perpendicular from the upper right corner.

"I extend the base line to the right.

"I label the two new points *e* and *f.*"

FIG. 2

With the help of this figure he then proceeds to the usual proof of the theorem that the area of a parallelogram is equal to the product of the base by the altitude, establishing the equality of certain lines and angles and the congruence of the pair of triangles. In each case he states the previously learned theorem, postulate, or axiom upon which the equality or congruence is based. Finally he concludes that it has been proved that the area of a parallelogram is equal to the base times the altitude.

"You will find what I have shown you in your textbook on page 62. Do the lesson at home, repeat it carefully so that you will know it well."

The teacher now gives a number of problems all of which require finding the areas of parallelograms of different sizes, sides and angles. This being a "good" class, the problems are all correctly solved. Before the end of the hour the teacher assigns ten more problems of this kind for homework.

2. At the next meeting of the class, one day later, I am there again.

The lesson begins with the teacher calling on a pupil to demonstrate how the area of a parallelogram is found. The pupil does it exactly. One sees that he has learned the problem. The teacher whispers to me: "And he is not the best of my pupils. Without doubt the others know it as well." A written quiz brings good results.

Most people would say, "This is an excellent class; the teaching goal has been reached." But observing the class I feel uneasy, I am troubled. "What have they learned?" I ask myself. "Have they done any thinking at all? Have they grasped the issue? Maybe all that they have done is little more than blind repetition. To be sure, they have solved promptly the various tasks the teacher has assigned, and so they have learned something of a general character, involving some abstraction. Not only were they able to repeat word for word what the teacher said, there was easy transfer as well. But—have they grasped the issue at all? How can I clarify it? What can I *do*?"

I ask the teacher whether he will allow me to put a question to the class. "With pleasure," he answers, clearly proud of his class.

I go to the board and draw this figure.

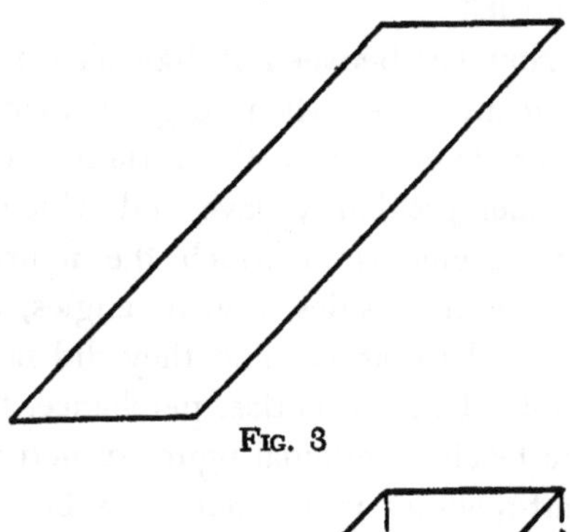

FIG. 3

Some are obviously taken aback.

One pupil raises his hand: "Teacher, we haven't had that yet."

Others are busy. They have copied the figure on paper, they draw the auxiliary lines as they were taught, dropping perpendiculars from the two upper corners and extending the base line. Then they look bewildered, perplexed.

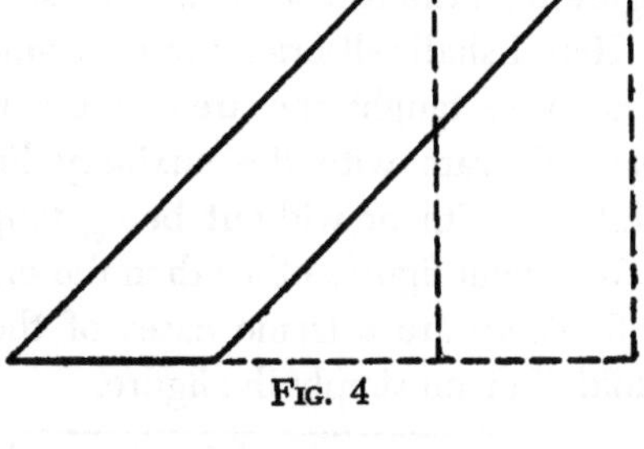

FIG. 4

Some do not look at all unhappy; they write firmly below their drawing: "The area is equal to the base times the altitude"—a correct subsumption, but perhaps an entirely blind one. When asked whether they can show it to be true in this case, they too become perplexed.[1]

[1] A boy from another class, observing these difficulties, whispers to me, "In

With still others it is entirely different. Their faces brighten, they smile and draw the following lines in the figure, or they turn their papers through 45°, and do it.

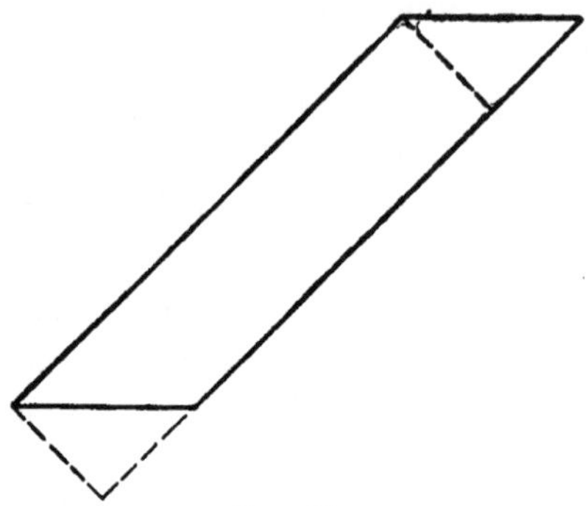

Fig. 5A

The teacher, observing that only a minority of the pupils has mastered the problem, says to me with some indignation: "You certainly gave them a queer figure. Naturally they are unable to deal with it."

Fig. 5B

Now just between us, haven't you too been thinking: "No wonder so many failed when he gave them a figure so unfamiliar!" But is it less familiar than the variations of the original figure which the teacher previously gave and which they solved? The teacher did give problems in which the figures varied greatly with regard to length of sides, size of angles, and size of areas. These were decided variations, and they did not appear at all difficult for the pupils. Did you notice, perchance, that my parallelogram is simply the teacher's original figure turned around? With regard to all the part-qualities it was not more but less different from the original figure than the teacher's variations.

Here I shall tell briefly of experimental proceedings with children who were taught the area of the rectangle, then the area of the parallelogram with the auxiliary lines and the result: base times altitude, with or without being taught the proof. They were then asked about figures other than the original figure.

3. There are extreme cases of thoughtless reactions, in which a child, if given simply the figure,

our class we have learned how to work this overlapping example. It's the teacher's fault. Why didn't he teach them how to do it in the overlapping diagram too?" To my surprise this special and complicated proof is sometimes taught from the start in the textbooks; really to understand it is not only difficult for children but it is also unnecessary for the solution.

FIG. 6

repeats what the teacher has said word for word like a blind slave, mumbles, "One perpendicular from the upper left corner," and draws it; then, "Another from the upper right corner," and draws it, "Extend the base line to the right," thereby getting this figure:

FIG. 7

4. On the other hand, it *does* happen that children even as young as six, who have learned nothing of geometry, after being shown briefly how to find the area of the rectangle, *do* produce the solution for the parallelogram in fine, original processes without being taught what to do. Some of these cases will be described in the third part of this chapter.

And it does happen that, having found or having been shown how to get the area of the parallelogram, children who are asked to find the area of the trapezoid, or of any of the following figures, are not at all helpless, but after some deliberation, sometimes with a little help, produce fine, genuine solutions of the kind that follow.

These are the tasks:

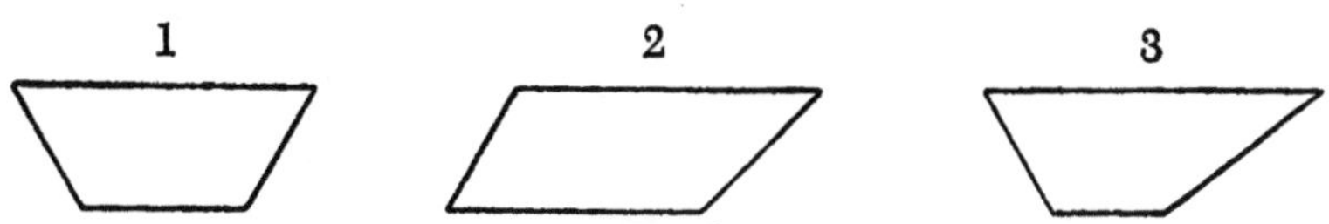

FIG. 8

In all these figures it is possible to solve the problem by changing the figures sensibly (A-responses) or to apply the learned operations, or some of them, blindly and unsuccessfully (B-responses).

A-responses:

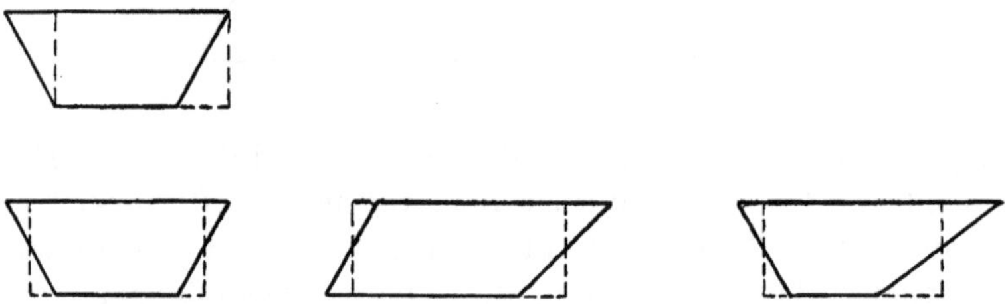

Fig. 8A

The subjects change the figures into rectangles by shifting the triangles. They will not give

B-responses:

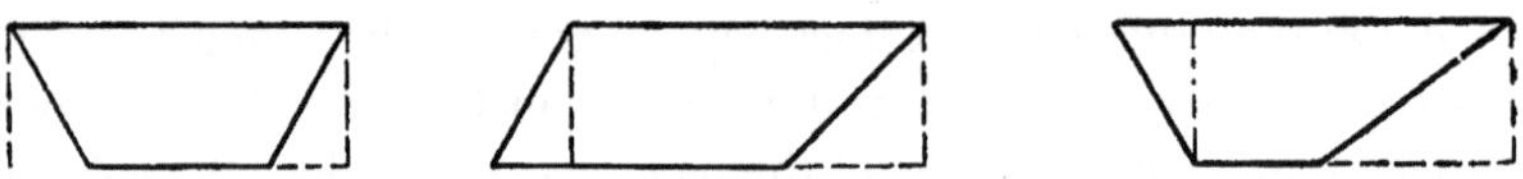

Fig. 8B

5. But others do give B-responses, or give A- and B-responses indiscriminately. Many pupils refuse to touch problems 1, 2, and 3 at all, with the comment: "How should we know? We haven't learned how to do those figures."

6. I then conducted experiments with children. I put before them single figures or pairs of A- and B-figures right after showing how the area of the parallelogram is to be found with the auxiliary lines.

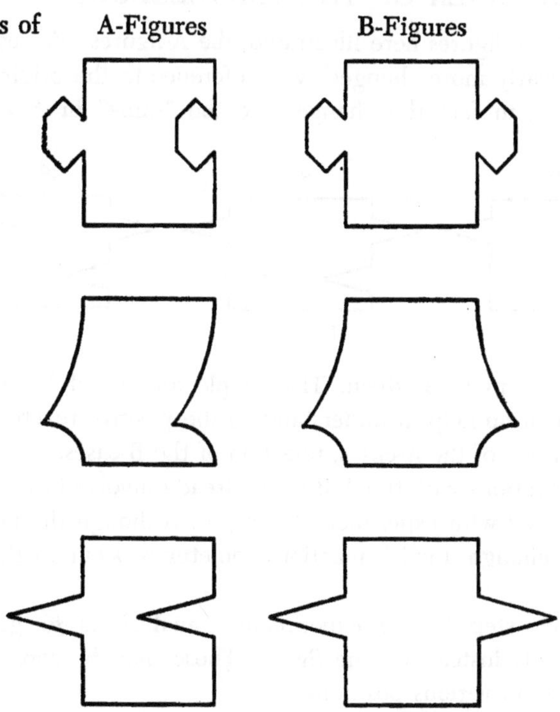

FIG. 9

In these pairs of figures, one member of the pair, the B-figure, can have no A-solution, whereas the A-solution is possible in the A-figure. For some children there seems to be no difference between the A- and B-figures. They are all new, "How should we know!" is their attitude. They either show no reaction or, if they do, they in no way differentiate between the A- and B-figures, draw some auxiliary lines and give blind answers.

But some consistently solve the A-problems and, sometimes after a brief period, reject the B-problems with: "This one I can't do; I don't know what the area is"; or even, "I don't know what the area of these little remainders is." By contrast the area of the remainders is generally not mentioned in the A-cases; or the child may say: "Of course I do not know the area of the little figures, but that does not matter as long as they are equal."

7. Also in the figures here illustrated, the A-figures, viewed piecemeal, are clearly more changed with reference to the original than the B-figures; in fact the changes are the "same" in A as in B,

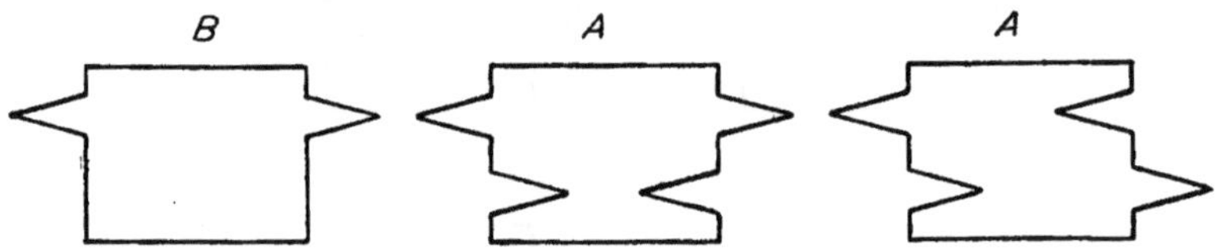

FIG. 10

only there are more of them. The simple answer of "familiarity" is obviously of no help in understanding the positive reactions—the straight solution of the A-cases, rejection of the B-cases.

Our observations with the A-B pairs already involved examples of how to proceed with experimental analysis. Although the task here seems easy enough, foolish reactions sometimes occur in the classroom.

8. Another step in the experimental analysis is to give two movable solids instead of one figure. These may be separated or adjacent, and in various positions:

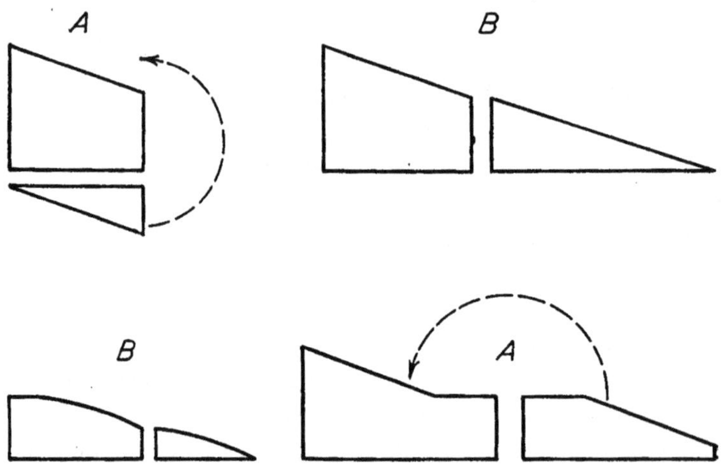

FIG. 11

Again foolish reactions are possible—and sometimes occur.

9. In order to become clear about the theoretical questions involved, it is sometimes helpful to consider extreme cases. Take the following foolish reaction:

A child is taught the proof of the theorem for the area of a parallelogram with the help of a figure drawn on graph paper. The auxiliary lines are drawn. The side *a* happens to be 5 inches, the segment *c* happens to be 3 inches long.

FIG. 12

The teacher says: "See! From each upper corner I drop a perpendicular of 4 inches; I extend the base line 3 inches to the right, you may measure it."

After a time another example is given, a parallelogram of different size. Suppose that a child were to proceed in a slavish way, distracted by an observer perhaps, or thinking of a game he was going to play, or wondering where his mother was at that moment; suppose a child were to repeat to himself, "Four inches down, three inches to the right," and were sheepishly to produce this figure:

FIG. 13

Then, asked if the goal, the area, had been reached, he replied: "... no," but could go no further for the time being. I myself have not found such a response, but it *may* happen. Similar things do happen in cases structurally less simple, as teachers know.

Clearly this would be an extreme B-reaction: copying what the teacher taught, blind to the context. Everyone sees what is wrong with it. But what is it, theoretically? One might say: "This child did not properly adjust what he had been taught to the new situation." But what does "properly" adjusting mean?

Or one might say: "Clearly this is a case of lack of 'generalization,'" and dismiss the matter as settled. Is it settled? What of foolish generalizations, which are no less generalizations for being foolish? What if a child were to generalize, for example, from the figure above (I have not found such a case): "The perpendiculars have to be one inch longer than the extension of the base line," or "The length of the perpendicular must be an even number," etc., and act accordingly?

To dismiss our problem with the term "generalization" does not help at all. Of course there is generalization here, but we have it in *both* kinds of cases. Often the emphasis on generalization does not answer the question; rather it covers up the problem.

10. What is the real issue in the A-B reactions, in the A-B cases? I had characteristic experiences: there are the intelligent reactions, the subject refusing to apply blindly to the B-problems what he has previously learned, giving the intelligent, genuine solution in the A-cases, altering the procedure as sense requires. And there are the blind reactions in which the subject is unable to deal with A- or B-cases, or applies operations foolishly.[2]

If a subject applies the method he has learned to a variation of the original problem, not realizing that it is out of place in the given instance, this is an indication that he has not understood in the first place, or that he has failed to realize what it is that matters in the changed problem. But if he deals appropriately and consistently with the A-cases even when there is more piecemeal dissimi-

[2] The foolish drawings in the examples on pages 18-19 one actually finds only comparatively rarely. Children with a spontaneous, natural attitude do not tend to behave that way. The habit of thoughtless repetition, as developed in certain schools by emphasizing blind drill, does seem favorable to responses of this kind; the same is true of situations in which a blind, piecemeal attitude has been created, e.g., by some distraction, by worry, or by personal difficulties. In schools that use drill the attitude often develops of responding to a new problem by just waiting to be told how to do it; if the pupil is asked to try it by himself, one often finds just passive refusal, with the comment, "We haven't had that."

That the psychologist felt uneasy in the class (see p. 16) meant that he felt the prevailing drill atmosphere. The behavior there described seems especially connected with the drill attitude, sticking blindly, as it does, to the words of the teacher—generally young children are not much bothered by the spatial positions of figures (cf. W. Stern, "Über verlagerte Raumformen," *Zeitschrift für angewandte Psychologie*, 1909, Vol. 2, pp. 498-526).

There are grownups who carry into their later life the acquired habit of blind, mechanical activity. It is astonishing to see how educated and otherwise intelligent people sometimes react in similar situations, especially in an "Einstellung" (see Chap. III, Sec. IV).

larity with the original, and if at the same time he refuses to apply the learned procedure to more similar B-variations, then it means that he has a real grasp of the problem. Thus A-B variations, systematically investigated, may furnish the basis for an "operational definition" of understanding. And in the course of experimental analysis the various structural factors may be studied by the A-B method.

What is the central difference between the two kinds of reactions to the variations? What, psychologically, is the issue? How does the subject find the A-reaction? What decides, in the mind of the pupil, between the A- and B-procedures?

First: one might say, "The difference is quite clear. The B-reactions do not lead to the correct solutions, while the A-reactions do." But this assertion states the problem; it does not solve it.

Second: "The degree of similarity to the original problem is decisive." No. Similarity does play a role, but what kind of similarity? Viewed in a piecemeal way the B-cases are often more similar to the original than the A-cases.

Third: is the matter explained by "generalization"? No. Of course generalization is involved in all these cases, but a foolish B-reaction, as previously stated, may involve as much generalization as an A-reaction. And so generalization in itself does not help. It would of course help if one were to speak of "properly selected generalization." But what are we to understand by this qualification? That it leads to the solution? This would again be like the first proposal.

Fourth: the situation remains unchanged if one asserts (correctly) that the various A-cases are characterized by grasping the essentials, by grasping what is really relevant. But what is that "grasping"? What are "essentials"? What determines what is and what is not essential? Only the result?

The theoretical proposals 2, 3, and 4 all fail to differentiate satisfactorily between the A- and B-reactions and cases, unless proposal 1 is implied, which does differentiate, but only on the basis of results. None of these proposals in itself brings psychological understanding.

I address myself to the reader: please think this over. Do not be satisfied with superficial solutions. I think you will see the answer if you face these cases directly. You may have it on the tip of your tongue without being able to express it in so many words. I shall not pursue the analysis here, but shall return to it later.

II

11. The psychologist, under the strong and vivid impression of the strange behavior of children in certain schools, makes another and more extreme attack on the problem.

As in the case cited, I have often been puzzled to see how some classes behave while learning. Frequently they follow obediently enough the steps of the proof which the teacher is showing them. These they repeat, learn. Onè has the impression that "learning" is going on—yes; thinking—perhaps a little; but real understanding—no.

Another procedure was tried for greater clarification.

What I am now going to say is somewhat unpleasant, ugly. But you see, the psychologist must sometimes, for theoretical reasons, use methods which are not in themselves so nice.

Instead of using the usual, sensible method of deriving the area of the parallelogram, the pupils are told: "In order to get the area of a parallelogram, one has to measure the sides—we will call them a and b; mark the point on the base line which is directly underneath the upper left corner, then measure the distance between the left end on the line and the point, which we call c. In our figure $a=5$ inches, $b=9$ inches, $c=3$ inches.

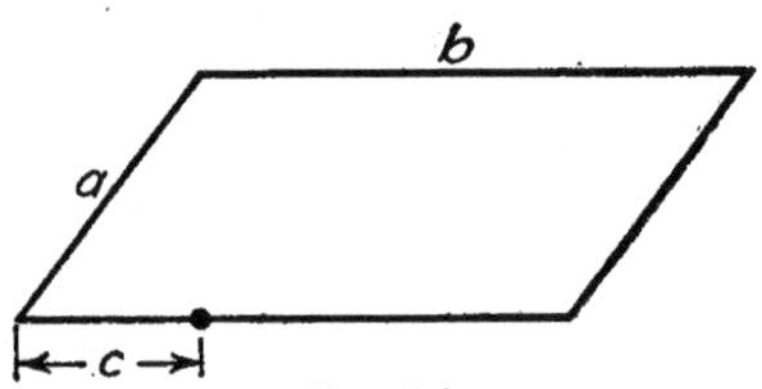

FIG. 14

"Now add a and c! $(a+c \ldots 5+3=8)$

"Subtract c from a! $(a-c \ldots 5-3=2)$

"Multiply the results! $(\ldots 8 \text{x} 2=16)$

"From the product extract the square root! You have learned that $(\ldots \sqrt{16}=4)$

"Multiply the result by b, and you have it . . . $(\ldots 4 \text{x} 9=36)$

"The formula for the area is $b\sqrt{(a+c)\ (a-c)}$."

This procedure is ugly and it would never occur to any reasonable teacher or mathematician. It takes the psychologist to introduce such a troublesome, unpleasant, and senseless method. But it does lead to a correct result.

Usually such a procedure seems strange and weird to children—one cannot help seeing that at certain points they draw back. In the end some of them look at the teacher with ill-concealed disdain. Others burst into peals of laughter.

It is a serious matter that in some schools one may *not* be able to observe any essential difference between the reactions of the pupils to this lesson and to a sensible method. When you find that children in a class swallow such a procedure piously, without any reaction, then look into the kind of education they are getting! I should think something must be wrong with it. And if you do make experiments of the kind here reported, I hope your classes laugh out loud, or at least that they are politely dismayed.

In instances of this kind it is sometimes touching to see the close concentration, the strong determination of the pupils to note every single step, the mumbling repetition of the teacher's words, the pride when they can reproduce exactly what they have learned, solve problems precisely in the way they have been taught. Teaching and learning is just that for many. The teacher has taught the "correct" procedure. The pupils have learned it; they can apply it in routine cases; that is all.

Let the reader consider whether he has not often learned things in school that way. Isn't it the way in which perhaps you have learned differential and integral calculus? Even theorems of plane and solid geometry? Of course you had good reason to feel that the teacher was teaching sensible, serious things you had to learn. But did you have the possibility of any other kind of learning, of really grasping? Could you do anything but put up with and submit to the teacher's demonstration, step by step, when you were unable to see why he did just this, then that? Could you help just following obediently as the steps dropped out of the blue?

I think you will agree that you could not. Nor would I be surprised if you were to add that surely it was the right way if the teacher did it that way, that probably there was no other. Or you may object: "One cannot compare this ugly example with the usual teaching in which the teacher sets forth reasonable things with their proofs."

You are perfectly right in this last remark. In our example the proof is lacking, an omission which, by the way, certain classes do not notice. And for a real decision we do need an example which includes a proof. This point will be further considered in ¶17.

12. But let us first get to the end of our story. I asked the class: "Are you sure that this result is really correct?" Most of the pupils were plainly dumfounded by the question, surprised that it should be asked. Their attitude was clearly: "How can you expect us to question the solution you have given us?" The question was strange to them, it touched the very essentials of what school, teaching, learning meant to them. No answer. The class was silent.

I changed my question, asked in a friendly way: "Could any of you show that the answer obtained in this way is really right?"

Little M. raised his hand. Looking quite shrewd, he said: "I know how one can prove it. It's very simple. We found that the area of this parallelogram is 36 square inches. Now I can cut the parallelogram out of tin, place it on one side of an accurate scale, and on the other I can place a rectangle which I know to be 36 square inches—I bet they'll balance."

"Yes, they may, but can you show that this would be the case generally?"

"Why, yes," he answered, "I can repeat it, if you like, with various parallelograms."

What this boy did is characteristic of many cases of thinking. He now has the *blind* procedure *plus* the proof by weight. That is all; he is satisfied. The proceeding, the so-called induction, is in itself a nice thing, often necessary, in some respects fundamental in modern empirical science. Nevertheless, to couple it contentedly with the blind, and therefore ugly procedure is not the solution for a real thinker, not the end. Modern science, while often based on induction, does not like to stop at induction. It goes on to search for a better understanding than induction alone provides. (See, for example, the development of the Mendeleev discovery.[3]) An

[3] Editors' note. Early in the nineteenth century the English chemist, William Prout, noticed that the atomic weights of the chemical elements were approximate multiples of the weight of the hydrogen atom, and he suspected hydrogen to be the *materia prima.* On the basis of this hypothesis, de Chaucourtois pro-

important instrument in its right place, induction is a start rather than an end in itself. But here it is not even justified as a start, being unnecessarily, arbitrarily blind to the issue.

13. To make the matter clearer, let us consider another example. The teacher demonstrates to the class how to get the area of the parallelogram by means of the usual auxiliary lines, shifting the triangle at the left to the right, showing at the end that the area equals the base times the altitude. In this example I suggested to the teacher that a parallelogram be used, one side of which, *a*, was 2.5 inches, and the other side, *b*, 5 inches. The altitude, *h*, was measured and found to be 1.5.

I then gave other problems, stating in each case the values of the sides *a* and *b*; the altitude was measured, and the area of each parallelogram was to be computed:

	a	*b*	altitude (measured)	area to be computed
1.	2.5	5	1.5	7.5
2.	2.0	10	1.2	12.0
3.	20.0	1⅓	16.0	21⅓
4.	15.0	1⅞	9.0	16⅞

The pupils worked at the problems, experiencing a certain amount of difficulty with the multiplication.

claimed in 1862 that properties of chemical elements were all governed by number. In 1871 Mendeleev published his famous periodic classification of chemical elements, a table in which all elements were arranged in eight vertical and seven horizontal rows. It enabled him to demonstrate that chemical elements changed their properties, especially their valence, in accordance with changes in their atomic weight. Atomic weight appeared to Mendeleev, therefore, to form the fundamental, essential characteristic of the elements. This was borne out by the fact that he was able to predict the discovery of unknown elements, necessary to fill the gaps in his table, through considerations based on the periodicity of properties and the regular increase in atomic weight of the chemical elements.

Though Mendeleev's classification was presented by him as a purely empirical generalization, it clearly pointed to the fundamental unity of matter.

In 1913, on the basis of the atom theories of Rutherford and Bohr, Moseley, a young British scientist, proved that it was the number of hydrogen atoms making up the atom of a given element, *atomic number*, rather than atomic weight that explained the chemical properties of elements—more exactly the number of protons and, therefore, the number of electrons.

Thus an empirical generalization became finally a deductive theory.

Suddenly a boy raised his hand. Looking somewhat superciliously at the others who had not yet finished, he burst out: "It's foolish to bother with multiplication and measuring the altitude. I've got a better method for finding the area—it's so simple. The area is $a+b$."

"Have you any idea why the area is equal to $a+b$?" I asked.

"I can prove it," he answered, "I counted it out in all the examples. Why bother with $b \cdot h$? The area equals $a+b$."

I then gave him the fifth problem: $a = 2.5$; $b = 5$; altitude $= 2$. The boy began to figure, became somewhat flustered, then said pleasantly: "Here adding the two does not give the area. I am sorry; it would have been so nice."

"*Would it?*" I asked.

This may serve as an example of a blind discovery, a blind induction. I dare say that no sensible mathematician likes to make such obviously senseless inductions. He will make them *only* if the matter under investigation is so obscure that he has no idea of a possible, sensible inner connection.

I may add that the real purpose of this "mean" experiment which succeeded, as you saw, was not simply to mislead. Visiting the class earlier, I had noticed that there was real danger of their dealing superficially with the method of induction. My purpose was to give these pupils—and their teacher—a striking experience of the hazards of this attitude.

One might say that the boy was wrong in his hypothesis simply because it was not universally true; that it was a generalization based on too few instances. But to do so is to miss the point. The proposed equality, area $= a + b$, is blind to the question of their inner relation, blind to the problem of how this can come about reasonably in even *one* case—for there is no inner relation between the area and "$a + b$."

14. I shall give a still simpler example. You ask a pupil:

1) $12 = 3$ times what? Answer: 4
2) $56 = 7$ times what? Answer: 8
3) $45 = 6$ times what?

Suppose the pupil were to say in answer to the third problem:

"Seven." And if you asked him for the reason, he were to answer: "Isn't it clear? The fourth digit is one higher than the third:

1)	12	3 4
2)	56	7 8
3)	45	6 7

Is the essential point here that the pupil has based his "hypothesis" on too few instances? No. The hypothesis *is* foolish: the lining up of the numbers in this way has nothing to do with the structure of the situation, is blind to its requirements, to the separation by the equality sign, to the meaning of the positions at the left, to the meaning of the multiplication sign at the right. It is blind to all these structural features, which set the requirements for a sensible solution or for a sensible hypothesis.

15. Here follow some additional examples of ugly procedures which lead to the correct answer. What is wrong here is not only that no proof is given but that none of the steps of the procedure has an understandable relation to the task.

How to get the area of a rectangle:

I

1) $a - b$
2) $\frac{1}{a}$
3) $\frac{1}{b}$
4) subtract 2) from 3)
5) divide 1) by the result obtained in 4)

$$\text{Area} = \frac{a - b}{\frac{1}{b} - \frac{1}{a}}$$

II

1) For $a + b$ substitute c
2) a^2
3) divide 2) by 1)
4) subtract it from a
5) multiply it by 1)

$$\text{Area} = \left(a - \frac{a^2}{a + b}\right) c = \left(a - \frac{a^2}{a + b}\right)(a + b)$$

16. I have chosen artificial examples in order to clarify the issue, but similar things happen without the intruding psychologist.

A child in school learns with accompanying drill the formulas for the perimeter, $2(a + b)$, and for the area, $a \cdot b$, of the rectangle.

After a time tasks are assigned which require computing the areas of rectangles in the course of larger problems. $2(a + b)$ comes to his mind and he uses it—all wrong without knowing it.

Or he tries to recall what the formula for the area was. He may even try to recall the page of the textbook on which the formula appears, and actually succeed in recalling it, but the formula just will not come to mind. He feels at a loss, looks at his neighbor's result, notes that the area was found to be 25—the sides, a and b, happen to be 10 and 2½, respectively. "I see!" he says to himself. "Now I remember how it's done: 10 + 2½ = 12½, times 2 = 25, $2(a + b)$," is relieved and busily solves the subsequent problems this way, again getting wrong results without knowing it. (Should it happen by chance that the next task has $a = 12$, $b = 2.4$, and that he looks for corroboration at the result his neighbor got, he will be reassured). It does not even occur to him to consider whether this formula could be at all reasonable here. On the other hand, if the pupil actually faced the problem, he might be able to reconstruct freely even a forgotten formula.

Now, is the decisive thing only that the boy got the wrong result, that his formula does not generally work? In order to bring out the issue sharply, think of a fantastic situation. The business of solving the problem could conceivably be done by a machine that cuts the rectangle into little squares. You put the rectangle into the slot; the machine does its work, the little squares drop out and are counted by you or by an adding device in the apparatus. Let us further suppose that as this machine works, it discards a certain number of the little squares, a number varying with the size of the rectangle; on the other hand, that four little squares are always added.[4] Such a

[4] In the formula $2(a + b)$ for the area, the area m disappears and the four shaded squares occur twice:

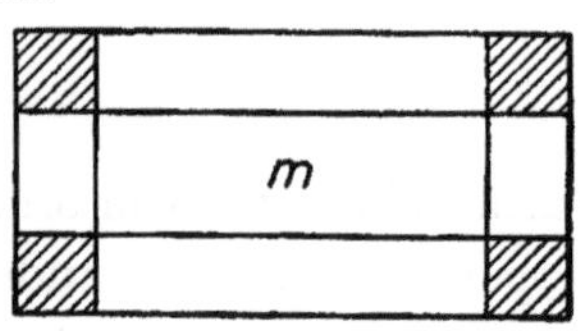

Fig. 15

machine can easily be constructed and would then by a general law invariably produce the result, $2(a + b)$.

The thinker urgently feels the need of looking into the machine to discover how this funny result comes about lawfully. If one could only open the machine and look inside! But suppose you could not, or even that there were no machine at all, that the same thing happened without the machine—miraculously, just in the course of cutting and counting . . .

You would have universality, a factual, constant formula, but the law expressed in the formula would still be ugly, blind to the issue, entirely ununderstandable.

17. Returning to our question, demonstrative *proof* was lacking in these ugly examples, and one might be of the opinion that this alone was the essential point. Let us consider what is necessary in order that a thought process be a good, sensible thought process. The usual answer seems to be:

> that a true solution of the problem is reached,
> that it is reached by logically correct operations,
> that the result is proved correct, universally correct.

Is that all? Is that an adequate expression of what we are facing in a real, sensible process?

Let us consider a proceeding which embraces all these features, and still is somehow ugly. Suppose I were to teach the area of the rectangle to a child who had heard nothing of geometry. First he is shown that the area of the square is a^2 . . . "*a* times *a*." He learns this and computes the areas of some squares of different sizes. He is then shown a rectangle, and the area of the rectangle is taught in the following way:

FIG. 16

1. First subtract *b* from *a*	$a - b$	$7 - 2 = 5$
2. Square the remainder	$(a - b)^2$	$5^2 = 25$

3. Square b and subtract it from the last result $(a - b)^2 - b^2$ $25 - 4 = 21$
4. Square a and subtract it from the last result $(a - b)^2 - b^2 - a^2$ $21 - 49 = -28$
5. Multiply it by -1 (make it positive) $a^2 + b^2 - (a - b)^2$ $+28$
6. Divide it by 2 ab 14

This is the area of the rectangle: $a \cdot b$ – which can be proved geometrically, as the figure indicates. The proof involves the demonstration that the two rectangles are congruent, and the subtraction of the overlapping b^2. Although it is somewhat intricate, it derives the general solution by logical necessity. This procedure is a little less ugly than the previous one, still it is ugly.

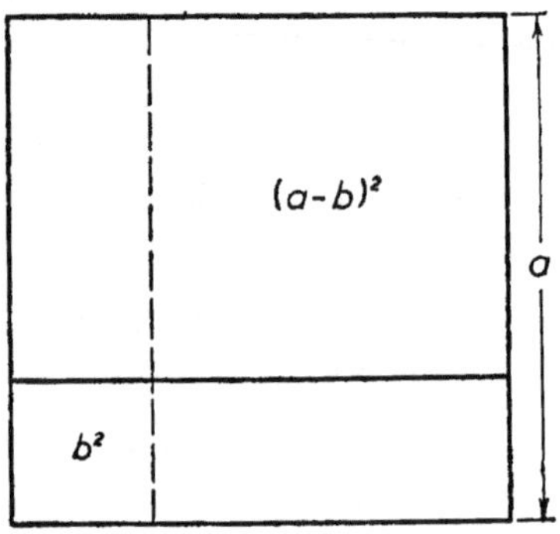

FIG. 17

Some reactions of children: "What grownups will do!" "Why not do it straight? It must be like the square—the number of little *squares in the base row* times the *number of rows*."

18. Let us look back now. Why are these various proceedings "ugly"? What is the decisive point?

1. Is it that the operations are not carried out correctly? No, they are in some examples.

2. Is it that they lack generality? No, the examples that are generally true are nevertheless ugly (cf. ¶11 and ¶15).

3. Is it that they lack demonstrative proof? No, not all the examples do.

If we look at the concrete proceedings in the ugly examples, see how the problem is attacked, how the individual steps in thought are related to the whole line of thinking, then the answer seems to be clear: I want to solve a problem; I face the problem situation; I want to see how I can clarify the issue in order to reach the solution. I try to see how the area is determined, how it is built up in this figure; I want to understand it. Instead, someone comes and tells me to do this or that, viz., something like $\frac{1}{a}$ or $\frac{1}{b}$ or $(a - b)$ or $(a - b)^2$, things which clearly have no inner relation to the issue, which go in another direction, a direction alien to the problem. Why do just this? I am told, "Just do it"; and then another step is added, again ununderstandable in its direction. The steps drop from the blue; their content, their direction, the whole process does not reasonably grow out of the inner requirements of the situation, appears arbitrary, blind to the issue of *how the area is built up* structurally out of the small units in *just this form.* In the end the steps do lead to a correct, or even proved answer. But the very result is seen in a way that gives no insight, no clarification. And this is true for all the examples—proof or no proof.

"Look here," the reader may say indignantly, "are you not demanding too much of human thinking?" I am not; happily there are thought processes which do not proceed so blindly.

19. The positive, productive course of thinking, as indicated in the reactions of children, is of an entirely different character. The question of the area in terms of the sum of the small unit squares is faced *in* the figure, with reference to *its characteristic form*: it is discovered that there are parallel rows which fit, each to the other, which are equal, containing the same number of small squares. Then the number of *squares in one* such row, determined by the length of one side, is multiplied by the number of *rows*, determined by the length of the other side. Here the essential thing is seeing the area structured in accordance with the characteristic form of the figure. None of the steps implied is in a direction blind to the issue, to the inner nature of the problem situation.

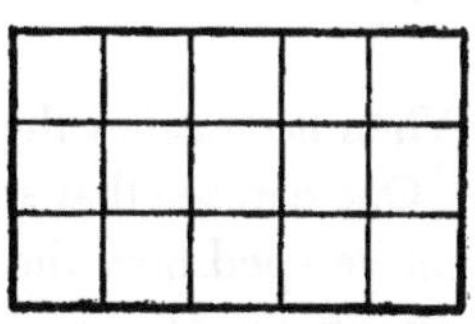

FIG. 18

And the very same result, area = $a \cdot b$, is psychologically not the same in sensible and in ugly procedures: "$a \cdot b$," as viewed in the sensible process, is not simply a "multiplication of two terms," for one of them *means* the number of *squares in one row*, the other the number of *rows*. The two terms in the multiplication have different structural and functional meaning and unless this is realized the formula, even the meaning of multiplication itself, cannot be understood.

20. I shall illustrate the last statement. A boy is shown a rectangle divided into the small square units. He is told that the total amount, the area, is equal to $a \cdot b$. Now, when he is given a number of different rectangles, he figures out the areas correctly by multiplying the sides. I ask him: "Are you sure this is correct?" And he replies, "Certainly, you taught it to me; but if you want me to, I'll count it out." And he starts counting sets of 5 squares, beginning at the base in the following way:

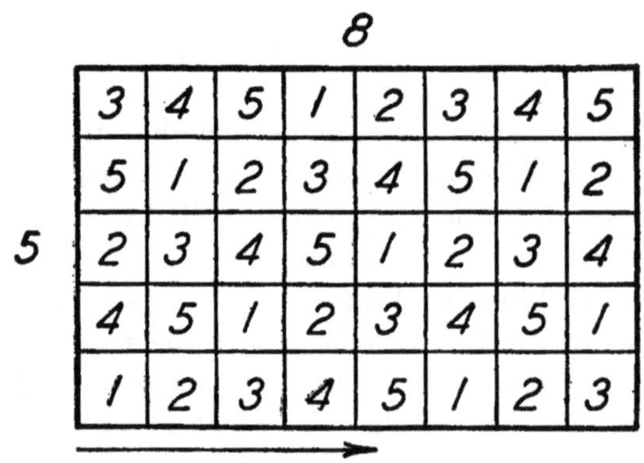

Fig. 19

When he reaches the end, he turns to me. "See? It's correct."

One can see that something essential is lacking here. This boy has not grasped how the area is built up by the row times the parallel repetitions. He hasn't used the fundamental structural feature that each row has the same number of squares. And so he has failed to get a basis for a sensible structural understanding of the area.

To put the matter differently, if the area is to be found by the kind of counting this boy did, the figure need not be a rectangle at all. Any other figure made up of continuous little squares would do as well. His procedure is blind to the structure of the figure in its inner relation to the multiplication.

Such structural understanding, or lack of it, plays a decisive role in transfer also. Here is a brief example: for the sake of experiment a child is shown how to find the area of a square. He masters it, applies it in various cases, and is then asked about the area of a rectangle. He sees no way of finding it. I ask: "Why don't you do it the same way as you did the square?" Hesitation, and then answer: "I can't do that . . . here the two sides are not equal."

But if the matter is really understood in the case of the square, the area is seen as the number of small squares in the base row times the number of rows, the transition offers no such difficulties. The fact of the equality of the two sides of the square is not disturbing then, being structurally peripheral, having no reasonable connection with the issue.

The transfer could also be blind. One could *without* this understanding simply believe blindly that the area of the rectangle too could be obtained by multiplication of the two sides. If one wishes to call both these cases generalization, one must realize that there is a basic difference between structurally blind, or senseless, and reasonable generalization.

21. Someone objects: "Why do you speak of grasping inner structure, inner requirements, implying that a grasp of the structural features in your example makes the thing reasonable? What of non-Euclidean situations? What if we choose other axioms for our geometry? What is reasonable in one system may be wrong in the other. Only as long as one presupposes the naïve, old-fashioned belief in the unique validity of the Euclidean axioms does what you say seem reasonable."

The objection is blind: it does not touch the issue. Non-Euclidean geometry has structural features of its own, to which again considerations of reasonableness apply within the new, enlarged framework. After having introduced the feature of space curvature, certain assertions of Euclidean geometry no longer fit, since they were made without taking into account the conditions introduced by the curvature and fit only the special case in which the curvature is zero.

To take a short illustration: a figure consisting of four "straight" lines and four right angles on the surface of a sphere *is* different from a plane rectangle also with respect to area; but here again you might get at the new area reasonably, grasping the inner structure, or you might derive it in an ugly way analogous to our ugly cases.

"Why do you speak of reasonableness in this context?" a logician asks. "Reasonableness is nothing but a question of old, formal logical consistency. Any theorem, any law—even your example [pp. 31-32] of the area of the rectangle as equal to $2(a + b)$ in the artificial world you described—is funny or unreasonable *only* in so far as it contradicts other laws, as it is not consistent with the axioms of its own system. That is all."

But this argument simply shifts the issue from theorems to axioms. If we envisaged changed axioms suited to just such structurally blind connections to give formal consistency, the result would be not just an ugly theorem, but the whole axiomatic system would be ugly.

To be sure, there are now trends in mathematics toward systems which would preclude reasonableness in the structural sense. Some believe it necessary to ignore all such reasonableness. A similar tendency appeared in the development of logic, the tendency to reduce logic to a game, governed by a sum of arbitrarily combined, piecemeal rules. As a division of labor such specialization is a meritorious enterprise, particularly with regard to criteria in the problems of rigid, logical validity. But if it is meant really to describe the whole business of logic, it reduces thinking to a poor thing, emptying it of features which play an important role in genuine productive processes. However the relation of the structural problems to formal logic and the theory of knowledge may be viewed, whether the structural problems should or should not be faced in logic, they constitute decisive issues in genuine, sensible productive processes.

The development of modern mathematics has moved in the direction of dispensing with all features of geometrical intuition as completely as possible. This was done with good reason, with an eye to questions of validity in an ideal, axiomatic system in which concrete theorems are obtained only by applying syllogistic and similar formal operations to the axioms. But this extremely well-justified interest should not be confused with the problems of understanding and of genuine productive processes. I have not found a really productive mathematician who did not feel the difference. Some have said: "This is not a logical or a mathematical question. It is a psychological question or, if you prefer, concerns the aesthetic side of the matter." To me it seems that such assertions view logic in too narrow a sense. The steps, the operations, in ugly procedures are not *reached* in a logical way—the direct procedure seems also the more logical. Indeed the difference between arbitrary, blind procedures and those which are not blind lies at the very heart of logic.

22. These have indeed been ugly, senseless examples, and the reader may have wondered why they were introduced. Common sense does in fact show them to be artificial and senseless; common sense feels the difference between these and really sensible proceedings. But it is necessary to focus on the obvious, in order to see the issue clearly and deal with it scientifically. Certain theoretical developments in logic, in the theory of knowledge, in psychology, have become blind to this issue or have tried to substitute criteria of a character that is blind to it.

Furthermore, things that we are inclined to take for granted as "obvious" need scientific clarification and work. I have used terms here that have the appearance of unfamiliarity, that are not too easy. But one must realize that the situation is fraught with problems. And it is not strange that it should be so. Whereas the operations of traditional logic have been richly and highly developed, those dealt with here are still largely unexplored. Gestalt theory is an attempt to develop them.

23. "The one point that suffices for the differentiation of what you call the ugly from the reasonable procedure," the logician interposes, "you have not mentioned. These examples do not look sensible simply because they employ *more* steps, a longer way around. You have forgotten the 'lex parsimoniae.'"

All the preceding solutions, it is true, do have more steps than the corresponding sensible ones. But one should not be deceived by this superficial feature; it is not the main point.

Do such "tricky" procedures always, and necessarily, employ more steps? Are they always more "difficult" than the corresponding sensible proceeding? No. In the area of the rectangle and the parallelogram the sensible proceeding is structurally too simple to allow of any shorter method, but mathematics offers such cases. Consider, e.g., the following problem.

What is the sum of the series

$$S = 1 + a + a^2 + a^3 + a^4 \ldots ? \ (a < 1)$$

The usual answer is:

1) Write down the equation	1. $S = 1 + a + a^2 + a^3 + a^4 \ldots\ldots$
2) Multiply both sides of the equation by a	2. $aS = a + a^2 + a^3 + a^4 + a^5$
3) Subtract the second equation from the first	3. $S - aS = 1$
4) Solve for S	4. $S = \dfrac{1}{1-a}$

That is correct; it is correctly derived, proved, and elegant in its brevity. A way to get real insight into the matter, sensibly to derive the formula is not nearly so easy; it involves difficult steps and many more. While compelled to agree to the correctness of the above proceeding, there are many who feel dissatisfied, tricked. The multiplication by a, together with the subtraction of one series from the other, gives the result; it does not give understanding of *how* the continuing series approaches this value in its growth.[5] Real understanding proceeds by considering what happens in the growth of the series and derives the law of this growth, leading to the limit.[6] Many do not bother really to understand. They are satisfied to have the result.[7]

There are theorems in mathematics for which we have at this time only "external" solutions because the problems are still too complicated for constructive understanding. Extreme are certain

[5] I cite an example of a subject's responses in one of my experiments: "Strange thing . . . multiplying by a . . . why this? Does it bring me nearer to the goal? . . . Subtraction—why? And now—in (3)—all that I know of the structure of S has disappeared! Was I not looking for the sum of such a growing series? I do not know more about it than before—only that it equals $\frac{1}{1-a}$. But why? How?"

[6] For a solution which, though requiring considerably more steps, does lead to a more direct understanding, see Appendix I.

[7] For the expert, to be sure, the usual procedure contains a sensible feature, based on the discovery that the series remains the same, with the exception of the first member, if you "shift" it, e.g., multiply it by a. Still, the procedure remains external to a real understanding of how the sum comes about.

cases of the so-called negative, indirect proof in which the principle of the excluded middle is used, showing that the opposite assumption is impossible, leads to contradictions, yet without any possibility of seeing how the positive solution comes about constructively. The famous mathematician, Brouwer, contemptuously called these indirect proofs "thinking with the spine." I am not here concerned as to whether he was right in demanding that results should be dropped which are derived only in such a way. What I want to emphasize is only that there is a striking difference between sensible solving, understanding the subject matter, and solving by external procedure.

III

24. Before reporting on some genuine thought processes of children in finding the area of the parallelogram, we pause to ask: What are the steps in a really sensible thought process solving the problem of the area of the rectangle? We shall enumerate briefly steps which seem to us essential on the basis of experiences with children and adults.

1) The problem is confronted: what *is* the area of the rectangle? I do not know. How can I get at it?

2) I feel there must be an *inner relation* between these two: the size of the area—the form of the rectangle. What is it? How can I get at it?

3) The area may be viewed as the sum of the small squares in the figure.[8] And the form? This is not *any* figure, not any heap of small squares in any form; I have to understand how the area is built up in this figure!

4) Are the little squares not *organized* in this figure, or *organizable* in a way that leads to a clear structural view of the total? Oh, yes. The figure is throughout equal in length, this has to do with the way the area is built up!

FIG. 20

[8] I here omit processes that start with variation of the size of the rectangle; introducing the small squares simplifies the picture. This procedure was some-

The parallel straight rows of small squares fit together vertically in mutual equality, thus closing the figure: I have rows equal in length throughout, which together form the complete figure.

5) I want the total; how *many rows* are there? I realize that the answer is indicated by the altitude, the side *a*. How long is *one row*? Obviously this is given by the length of the base, *b*.

6) So I have to multiply *b* by *a*! (This is not multiplication of two items equal in rank: their characteristic functional difference is basic for the step.)

In this structurization of the rectangle the question of the area becomes clear. The structure obtained is seen comprehensively and transparently. The solution is reached[9] in the realization of the inner structural relation between area and form.

25. I do not mean to assert that these enumerated steps are always separately formulated in actual thinking processes![10] They are mostly in global interplay; yet I think they are needed for any real understanding of the issue.

They involve a number of operations, a number of features, which were not really seen or dealt with in the approaches of traditional logic and associationism.

1) There is *grouping, reorganization, structurization*, operations of dividing into sub-wholes and still seeing these sub-wholes together, with clear reference to the whole figure and in view of the specific problem at issue.

These are operations performed not just in any way; we have here not *any* grouping or organization, even though in fact many different kinds are possible; the steps are conceived and carried out *in*

---

times found by children; sometimes the experimenter gave the rectangle as an arrangement of cube blocks or drew the lines from the beginning; in these cases essential steps had still to be taken by the child.

[9] In step 4), instead of the horizontal rows, the vertical rows could have been chosen. But one should not, in thinking, mix up these two ways. If the child does, the difference between "number of rows" and "length of a row" easily becomes blurred; it is advisable therefore to start with a rectangle the sides of which are markedly different. Step 5) is easily clear if the sides are simple multiples of the side of the measuring square; if not, another step is involved, viz., to decrease the measuring square. In 5) and 6) multiplication comes in. This is by no means simply or necessarily recall of the operation learned in arithmetic. It may be just the opposite: the idea of multiplying or the meaning of multiplication may arise in just such a context.

[10] I would not advise predigesting each of these steps for the child in teaching. But it is sometimes helpful to ask a question in one of the indicated directions.

accordance with the whole-qualities of the figure and with the aim of getting at a *clear structure* of the area.

There is involved the realization of how parts (sub-wholes) *fit* together and *complete* the area; the realization of the *inner relatedness* between their fitting together and the whole-features of the figure, such as straightness of the lines, etc.

2) The process starts with the desire to get at the inner relatedness of form and size. This is not a search for just any relation which would connect them, but for the nature of their intrinsic interdependence.

Here some people proceed to introduce changes, observing, studying what a change (e.g., in the width of the figure) does to its form and size; and in this way they get at the kind of inner relatedness involved.

3) Outstanding relations of this kind—sensible with regard to the inner structural nature of the given situation—which we will call ρ-relations, play a large role here:

Equal, straight, parallel rows fitting together:	form of the rectangle involving straightness of lines, not for instance a structure like 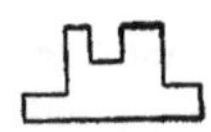
Number of rows:	length of one side
Number of squares in a row:	length of the other side
Multiplication:	completing the structure

4) There is the feature of the functional meaning of parts, e.g., the characteristically different meaning of the two terms entering into the multiplication, a feature decisive for the productive solution and for any real understanding of the formula.

5) The entire process is *one consistent line of thinking*. It is not an and-sum of aggregated, piecemeal operations. No step is arbitrary, ununderstood in its function. On the contrary, each step is taken surveying the whole situation. No step appears here that is of the character of $a - b, \frac{1}{a}$ or $(a - b)^2$ in our ugly examples.

The essential features of the operations mentioned are basically different in character from the operations of traditional logic and associationism, which are blind to the view of the whole, to the structural requirements which actually bring the operations into existence.

I hope the reader realizes the dramatic and consistent character, the beautiful clarity of such a process and its stark difference from processes in which the operations drop in senselessly.

26. By contrast, a description of the process in terms of traditional logic only, or of associationism, looks poor indeed.

Here I want to mention one point with regard to these approaches. In traditional logic the feature of universality is basic: in the concepts, in the judgments, one looks for common qualities in many or all cases (in this instance for the common qualities in many rectangles). Similarly in associationism the question of many cases, of many repetitions showing some constant connection, is fundamental. In line with this, the ugliness of our cases of induction was viewed as due to their lack of general validity. But the issues of sensible structurization, organization, of fitting of items into each other, of completion, etc., are not necessarily connected with thinking of other cases; they may be envisaged, realized in the single concrete case, viewing *it* structurally, sensibly. This does not of course give assurance of universality in factual questions; but it often leads to reasonable understanding and to genuine discovery of essential features, in contrast to performing operations on the basis of blindly generalized features common to many or all cases. And it may imply structurally sensible transposability of items (see ¶4), leading to sensible generalization and universality. But the steps do not necessarily come into existence by viewing many cases and stating what may be factually common in them.

27. Finding that the usual concepts did not suffice, some theorists arrived at the formulation that using *relations* makes thinking productive. Of course realizing relations does play an important role in thinking, but the assertion in itself is still blind to the main issue, is not the solution. For the difficulties that we encountered with regard to elements recur similarly with relations. To realize *any* relations, even if they are correct, is not decisive; what is decisive is that they must be the relations structurally required in view of the whole, arising, conceived, used as parts in their function in the structure. And this holds equally for all operations of traditional

logic and associationism, such as generalization, abstraction, etc., when used in genuine thought processes.

Incidentally, ugly and unsuccessful proceedings involve relations no less than do productive ones.

28. In line with another modern approach one might reflect: "The difference you emphasize between the ugly and the good cases is in fact simple and means only this: in what you call the ugly cases, means are employed, steps, operations, concerning which I do not know in advance that they will help in the end. Whereas in the procedure you call sensible, I do know this on the basis of past experience. I know beforehand, for example, that where an amount is divided into equal parts I can use the learned technique of multiplication. I am using means here which have been associated with their effects by previous drill. Association brings about the recall."

There is nothing to be said against the first part of this formulation, that in the ugly cases means are employed concerning which I do not know in advance whether they will help. But the second part of the formulation is inadequate: first, it ignores the operations of fitting, grouping, etc., and their characteristic features; secondly, knowing some constant connection between means and ends and employing it is not the issue here. "Knowing" is an ambiguous term. To know a blind connection, such as the connection between the switch and the light, is very different from realizing or discovering the inner relatedness between means and ends, their structural fitting in our case (see ¶38). This difference *is* of importance in this context precisely with regard to the coming into existence of the sensible, productive processes.

And the formulation with regard to the recall of multiplication learned by drill does not fit our sensible cases. For multiplication and its meaning are sometimes initially conceived through realization of structural needs in just such tasks. And even if the technique of multiplication has been learned before and is recalled now, the essential thing is what is known and what is recalled: some drilled operation blindly applied, or just the operation structurally required, recalled and used for that reason and not on the basis of some peripheral association (such as having done a lot of multiplication just the day before or heard the word "area" connected with the word "multiplication").

29. Multiplication is not simply a matter of being taught operations in terms of associations, connections between numbers. If sensible, it rests on a structural discovery or realization that is necessary even in its application. It is unfortunately true that many children are drilled in multiplication, can react instantaneously but have no idea where to apply it.[11]

30. Now I shall tell what happened when I put the problem of the area of the *parallelogram* to subjects, especially children, after having briefly shown how the area of the rectangle is found, saying nothing further, giving no help, simply waiting for what they would say or do. There were grownups of all types, students who showed by their reactions that they had entirely forgotten this theorem, and children who had never heard of geometry, even children as young as five.

There are different types of reactions.

First type. No reaction at all.

Or someone says, "Whew! mathematics!" and dismisses the problem with, "I don't like mathematics."

Some subjects simply wait politely for what is to come or ask, "What else?"

Others say, "I don't know; that is something I have not learned." Or, "I learned that in school but I have completely forgotten it,"

[11] I used to ask a child—there were often guests in the house—"How many men and how many women are there at the table? How many are there altogether?" I asked this question often, when the child was six, then seven, then eight years old. In school she was good in arithmetic. If you asked her to multiply, say 6 × 2, she would answer correctly in "instantaneous reaction." But in this case, even when there were four men on one side of the table and four women on the opposite side, or if the men and women were sitting in pairs, she would proceed to count tediously: "One, two, three, four men; one, two, three, four women." Not until she was eight and a half years old did she have the idea, after having counted the men, of saying, "And there are just as many women," or "One, two, three, four pairs." And she is an intelligent child. Only she did not realize the bearing of the grouping on the amount, habituated as she was to counting one by one.

On the other hand, at the age of six, in more difficult but structurally clearer situations, her performances were astounding by comparison. I asked her, as I did many children, to figure out, in imagination, the number of sides and corners in a cube of sugar, then in the pyramid and in the double pyramid. She was able to find the answer structurally, and to apply it generally to the pyramid and double pyramid, even to a pyramid which had 3 × 7 lines, although she was not able to count to, or even say, 21.

and that is all. Some show indignation: "How do you expect me to be able to do that?" To which I reply, "Why not try it?"

Second type. Others search their memory intensively, some even frantically, to see if they can recall anything that might be of help. They search blindly for some scraps of knowledge that might apply.

Some ask, "Could I ask my older brother? He surely knows." Or: "Could I look for it in a geometry book?" Which is certainly one way of solving problems.

Third type. Some start making speeches. They talk around the problem, telling of analogous situations. Or they classify it in some way, apply general terms, perform some subsumptions, or engage in aimless trials.

Fourth type. But in a number of cases one can observe real thinking at work—in drawings, in remarks, in thinking out loud.

1) "Here is this figure—how can I get at the size of the area? I see no possibility. The area just in this form?"

2) "Something has to be done. I have to change something, change it in a way that would lead me to see the area clearly. Something is wrong." At this stage some children produce Fig. 21. In such cases I add: "It would be nice to be able to compare the size of the area of the parallelogram with the area of the rectangle." The child is helpless, then starts anew.

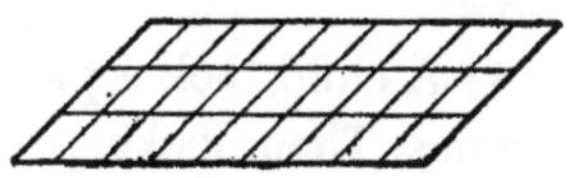

FIG. 21

There were other cases in which the child said: "I have to get rid of the trouble. This figure cannot be divided into little squares."

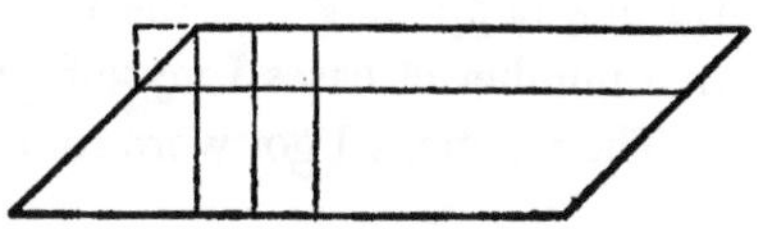

FIG. 22

3) Here one child said suddenly: "Could I have a folding ruler?" I fetched one. The child made a parallelogram of it, then turned it into a rectangle.

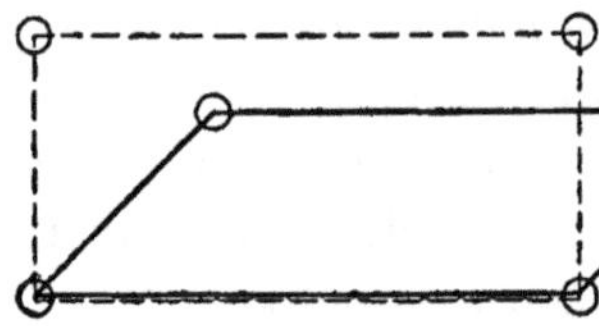

FIG. 23

I enjoyed this. I asked: "Are you sure this is correct?" "Sure," he said. It was only after considerable difficulty that I succeeded in making him doubt his method—by using an appropriate drawing such as

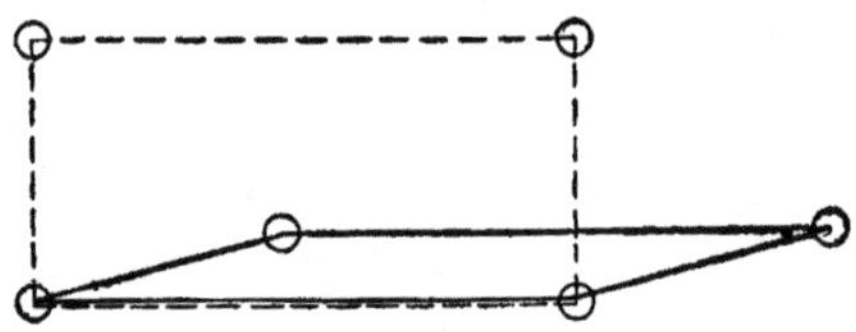

FIG. 24

Here he said at once: "The rectangle is much bigger—it doesn't work . . ."

4) A child took a piece of paper and cut out two equal parallelograms. Then, looking happy, he put the two together in this way.

FIG. 25

But he did not know how to proceed further.

This was in itself a fine happening (cf. the ring solution, p. 49). I may remark that in a number of cases I myself gave the child two samples of the figure. The reactions I got were sometimes as follows:

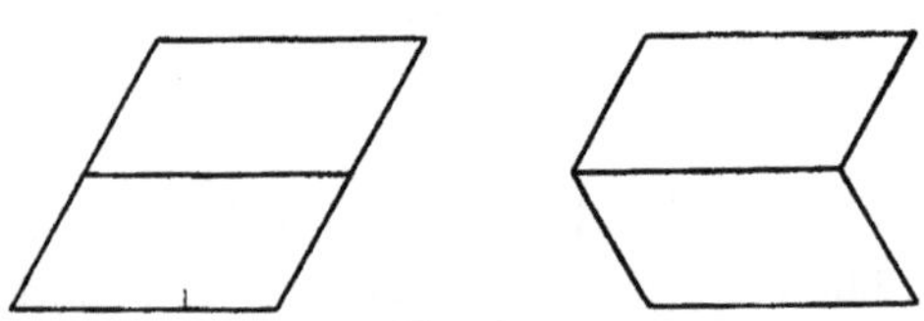

FIG. 26

Some indulged in several trials of this kind, even putting one figure above the other congruently or conversely. Such help seems to be effective only under certain conditions. Which ones?

31. But there were cases in which the thinking went straight ahead. Some children reached the solution with little or no help in a genuine, sensible, direct way. Sometimes, after strained concentration, a face brightened at the critical moment. It is wonderful to observe the beautiful transformation from blindness to seeing the point!

First I shall report what happened with a 5½-year-old child to whom I gave no help at all for the parallelogram. Given the parallelogram problem, after she had been shown briefly how to get at the area of the rectangle, she said, "I certainly don't know how to do *that.*" Then after a moment of silence: "This is *no good here,*" pointing to the region at the left end; "and *no good here,*" pointing to the region at the right.

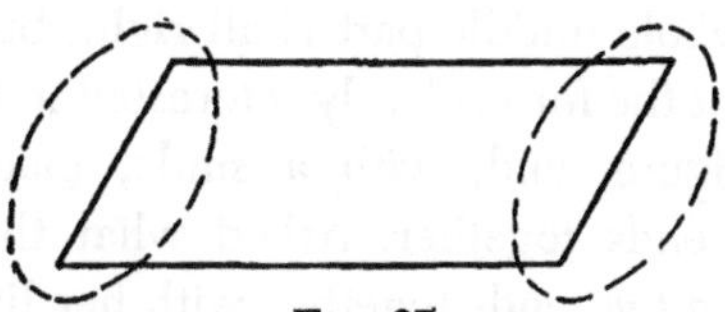

FIG. 27

"It's troublesome, here and there." Hesitatingly she said: "I could make it right here . . . but . . ." Suddenly she cried out, "May I have a scissors? What is bad there is just what is needed here. It fits." She took the scissors, cut vertically, and placed the left end at the right.

Another child proceeded in a similar way to cut off the triangle.

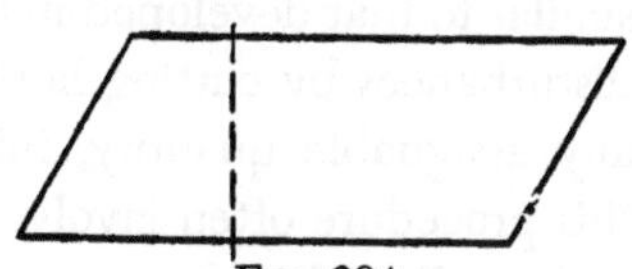

FIG. 28A

FIG. 28B

In several cases the procedure ran this way:

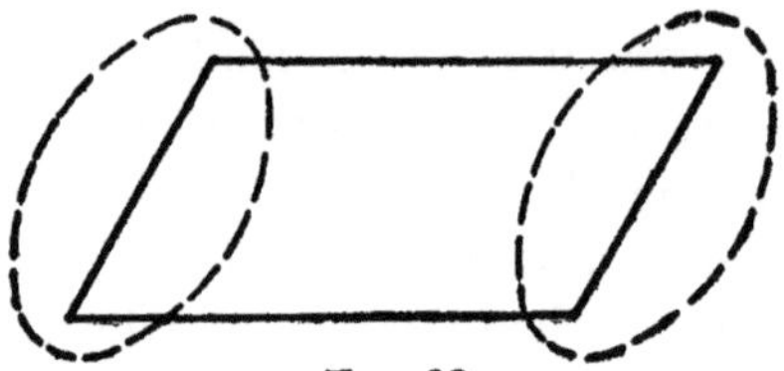

FIG. 29

1) "Disturbance"	"Disturbance also"
2) "Too much here"	"Too much here"
3) ————————	"No! This needs over here at the right just what is too much at the left,"

and she put the left end "in order." Then, looking at the other end, she tried to do the same thing there, but changed suddenly from seeing it as "too much" to seeing it as "gap."

There were other ways. A child to whom I had given the parallelogram, a long one cut out of paper (for the previous examples also, a long parallelogram seemed to be better), remarked in the beginning, "The whole middle part is all right, but the ends—" She continued to look at the form, clearly interested in the ends, suddenly took the paper figure, and, with a smile, made it into a ring, bringing the two ends together. Asked what this meant, she answered, holding the two ends together with her little fingers: "Why, I can cut it now, this way" and indicated a vertical somewhere in the middle, "Then it is all right."

Partially different procedures appear also, but I did not find any similar to that developed in modern mathematics, of diminishing the disturbances by cutting horizontal rows with altitudes smaller than any assignable quantity, infinitesimally small. Even for grownups this procedure often involves difficulties. Cutting in rows of smaller and smaller altitude, proposed to children of about twelve and to grownups, brought funny reactions. After being told the whole story some were still puzzled, thought it "insincere" even when shown that if the rows are shifted properly horizontally, the whole figure becomes more and more "similar" to the rectangle. It involves the transition to the concept of infinitely small and to the procedure of reaching a limit. This method became possible historically only after a long development oriented in a much broader region of geometry.

32. What are the operations, the steps in the procedure?

We saw that in such genuine, positive processes as those just described, there are again factors and operations similar to those mentioned in the discussion of the rectangle: regrouping with regard to the whole, reorganization, fitting; factors of inner relatedness and of inner requirements are discovered, realized, and followed up. The steps were taken, the operations were clearly done in view of the whole figure and of the whole situation. They arose by virtue of their part-function, not by blind recall or blind trial; their content, their direction, their application grew out of the requirements of the problem. Such a process is not just a sum of several steps, not an aggregate of several operations, but the growth of one line of thinking out of the gaps[12] in the situation, out of the structural troubles and the desire to remedy them, to straighten out what is bad, to get at the good inner relatedness. It is not a process that moves from pieces to an aggregate, from below to above, but from above to below, from the nature of the structural trouble to the concrete steps.

As far as we can see there is no blind trial and error in the good cases. And if it does occur, it is quickly dismissed. I have not in such processes seen really foolish, blind operations. I have not, for example, seen a case in which the triangle at the left is cut off and placed on the other side in this foolish position: I did not even find cases in which the trouble regions, having been recognized, were viewed in terms of the four angles individually.

FIG. 30A

FIG. 30B

33. It would be possible, of course, to get the externals of this procedure, even the solution, by senseless drill. Let us see clearly and honestly what this implies as a general, theoretical approach.

[12] Initially there is the gap concerning the size of the area. Tne desire is to fill this gap, to realize in the context of the other characterization what the size is, how this quality of size is structurally determined in the context. Especially when the parallelogram is a long one, the first step is clear: the middle region yields a transparently clear determination of the size—as in the rectangle; the ends appear as trouble regions, clarified in the process then by "making the ends good, too," which is based on the further realization of a second kind of gap: the one end viewed not as disturbance to be cut off but as needing completion.

Taking the extreme case, it would be possible to "teach" the procedure without even stating the problem first; the teacher draws the diagram. The pupils repeat "one auxiliary line," perhaps twenty times, in this way building up a connection with many reinforcements, with much recency. Then they continue similarly with another auxiliary line, "connecting" it with the figure, and so on, in this way reaching the end, the finished result. This *would* be at least a possible procedure, in line with association theory. I myself have not done such experiments. But I should think that even a positive result thus produced in blind pieces would differ strongly from the good cases with regard to future consequences, such as forgetting or application.

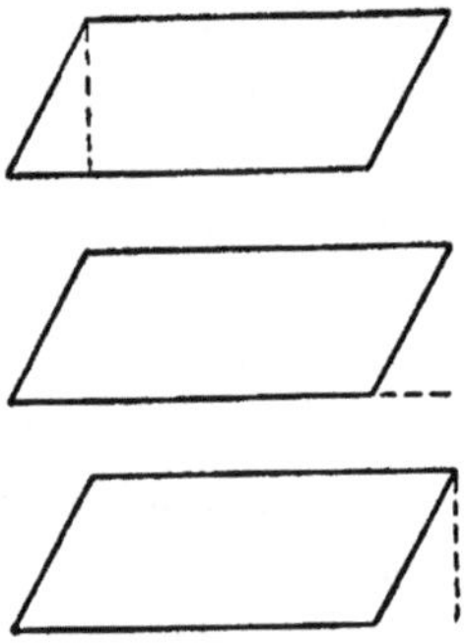

Fig. 31

Of course these remarks are excessively simplified as regards the theoretical situation. In order to give an exhaustive picture one would need to go into a discussion of all the auxiliary hypotheses developed in the association approach, which tries to base all sensible processes on an aggregate of mechanical, blind connections. What has been said here may serve merely to give a hint of the fundamental problems involved.

34. It was mentioned above that at times the pupil concentrates first on, say, the left end of the parallelogram, deals with the disturbance by cutting it off, then finds his way by dealing with the right end as a gap to be filled, as required by the felt disturbance there *and* the damage done in leaving a remainder at the left.

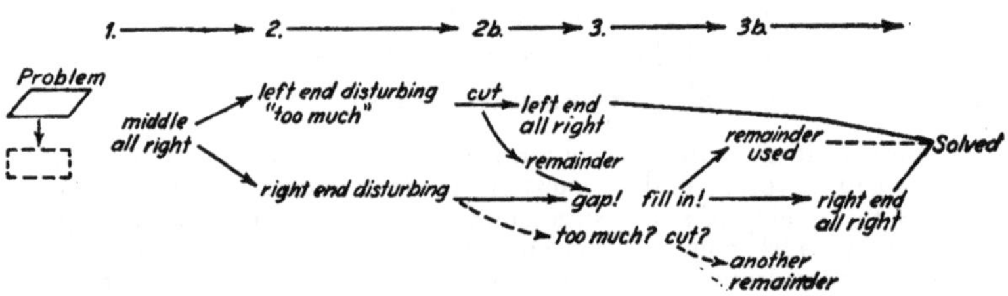

Such a sequence does not seem to portray adequately what happens in other cases in which the subject envisages the two directions, the correction of the two ends in one comprehensive view of the figure: what is disturbing, excessive, on the left is seen as just needed at the right.

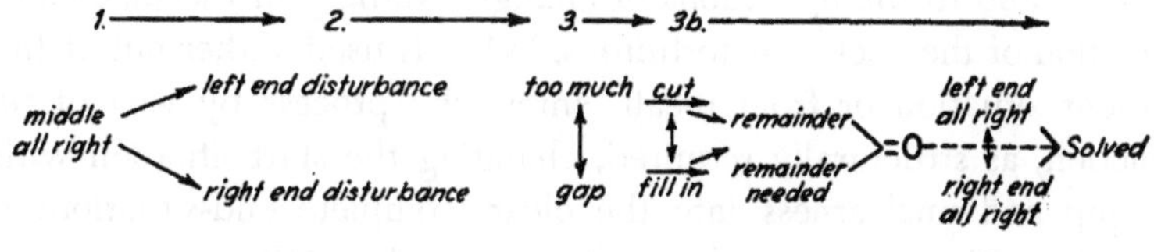

Both directions are here envisaged and dealt with as fitting together, as requiring each other.

This is still clearer in the ring solution: the two ends are seen as fitting together, as requiring to be brought together to make all trouble disappear. There is no real functional difference between them, they are seen as having the same disturbance, to be got rid of in the same way, by mutually completing each other.

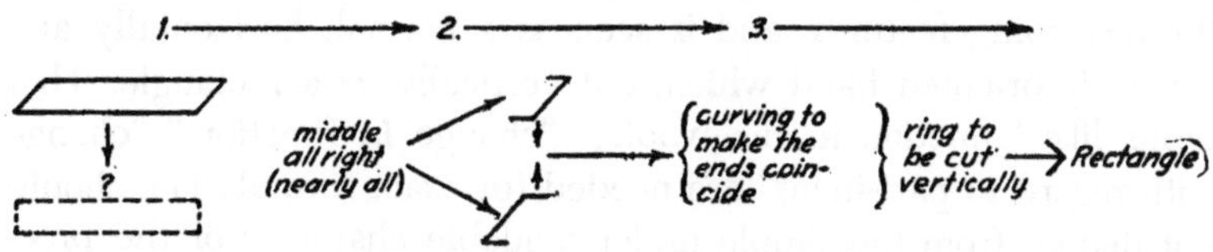

The solution by cutting in the middle and shifting is often similar:

↗ get proper rectangular ends by cutting somewhere vertically,
↘ get rid of the disturbing ends by fitting them together (shift).

If one has vividly experienced such cases, one feels that the great dangers to the development of such fine processes are mainly blind recall, blind application of something learned, a piecemeal painstaking attitude, a failure to view the whole situation on its merits, in its structure and in its structural requirements. It seems to me, although I do not have sufficient quantitative data on this point, that the ability to produce such fine, genuine processes often decreases considerably in school children when they become accustomed to drill.

What the diagrams illustrate is the line of the vectors in the process.

Dynamically the essentials in such thought processes seem to be briefly as follows: facing the problem, vectors arise in connection with and determined by the structural features, the gap, the incompleteness in the situation, tending to concretization of the trouble regions, and to the operations of change. Nothing in the place and direction of the vectors is fortuitous. What is used, either out of the present situation or from recall, enters the process by way of its function, as structurally required, changing the start-situation with its gap and unclearness into the clear, complete end-situation: a good transition from a bad gestalt to a good gestalt.

My description appears highly complex because I have described the various steps implied, separately and successively, and because I had to employ formal terms which are not habitual in the traditional approaches. But does it look so complex, for example, in the case of the ring in which the essence of the procedure is simply that the disturbing oblique sides lose their character of boundary lines by closing the figure, disappearing in the plane? The ring has lost the disturbing features, and is seen as a normal, horizontally and vertically oriented band which, cut vertically, *is* a rectangle. That terms like "function in the whole," "change in function," "change with regard to part-items" are needed for exact formulation, should not detract from the simple understandable character of the process.

35. Here I shall not trouble the reader with a thorough structural analysis of these processes. I shall give only an idea of some of the items formally involved in them.

If in such a proceeding the three auxiliary lines are produced, they come into the picture *not* as "this line drawn vertically from the left upper corner, a second from the right upper corner, and the third prolonging the base to the right," later perhaps acquiring some value, some meaning. But they come into existence from above out of the functional requirements, in their role as parts. And in the process parts of the figure change their functional meaning:

1) The auxiliary line at the left originates

a) as the rectified, properly made *end* of the rectangle at the left;

b) and appears at the same time not as some one vertical but as part of the triangle;

c) as such, it is transferred, shifted to the right, becoming the proper end of the rectangle at the right.

a) and b) already involve a double function[13] of the line, closing the triangle and forming the left end of the rectangle. In c) the line is shifted to the right with the whole triangle, functioning there as the proper right end of the rectangle.

The second perpendicular is again not just any line dropped from a corner, but comes into existence as the proper end of the rectangle, being the side of the triangle that was lacking.

And the prolongation of the base line comes into existence not as just any prolongation of the line but as part of the needed triangle, completing the rectangle base.

The three lines come into existence not as lines but as boundaries; the figures, not the lines, are the issue—the parallelogram, the rectangle, the triangle; the lines come in only as parts of them.

2) What happens to the original lines of the figure? Some subjects describe it vividly. The figure is viewed first as a parallelogram with its two horizontals in oblique reference to each other.

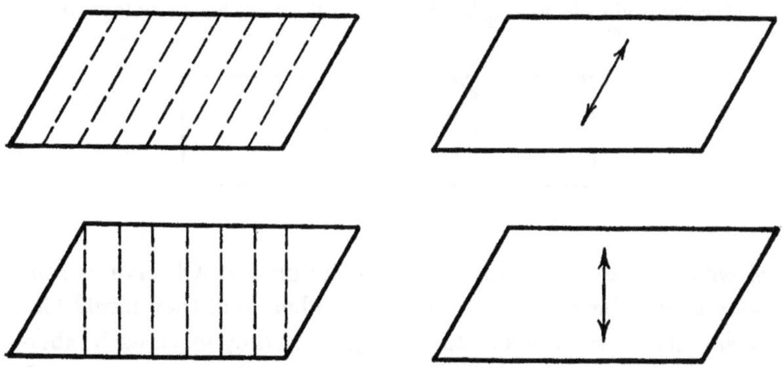

FIG. 32

But then the horizontals acquire a vertical reference to each other. The left end of the base is no longer co-ordinated with the left end

[13] M. Wertheimer, "Untersuchungen zur Lehre von der Gestalt," *Psychologische Forschung*, 1923, Vol. IV, pp. 301-350; see also W. D. Ellis, op. cit., selection 5. H. Kopfermann, "Psychologische Untersuchungen über die Wirkung zweidimensionaler Darstellungen körperlicher Gebilde," *Psychologische Forschung*, 1930, Vol. XII, pp. 293-364.

of the upper horizontal but is separated, to be taken away as the base of the triangle. The part of the base at the right appears unfinished, as lacking its proper end.

The two oblique sides become disturbing: "The ends of the figure should not look like that"; a vector arises not to have them as boundary lines; in the shift of the triangle, they jump into identity, are no longer two but one, and this one is no longer a boundary line, is now in fact structurally unessential.

Similarly in the case of the first solution (p. 48) and of the ring solution: the introduced vertical line is conceived, comes into existence with the double function of becoming the proper end of the rectangle at the left and the proper end of the rectangle at the right. (Real understanding of the line involves this splitting into the two functional items.) And the oblique lines fall into identity, disappear structurally.

Changes of a similar character can be studied in perception. Both the structure of the events and the strength of the forces involved are comparable in this region.

Here is a simple example:[14] the two figures shown below are cut

Fig. 33

of wood or cardboard, black on a white ground. Observe as somebody moves them slowly toward each other. Will they meet? Close each other? It is quite a sudden thing when they reach each other—fit—and the jagged sides disappear in one homogeneous, undisturbed rectangle.[15] And what takes place in observers if near the end the quiet, slow movement in the horizontal suddenly changes its direction slightly? Some people actually jump to correct it, to make the parts fit.

[14] Cf. M. Wertheimer, "Zu dem Problem der Unterscheidung von Einzelinhalt und Teil," *Zeitschrift für Psychologie*, 1933, Vol. 129, pp. 353-357.

[15] Compare also the square-sets in Chap. III, pp. 106-107.

Similarly in our parallelogram task: a child pondering on the task gets the idea of cutting off the triangle at the left; cuts it; you take the triangle to shift it to the right; what reaction do you find in children if you leave the triangle in these positions?

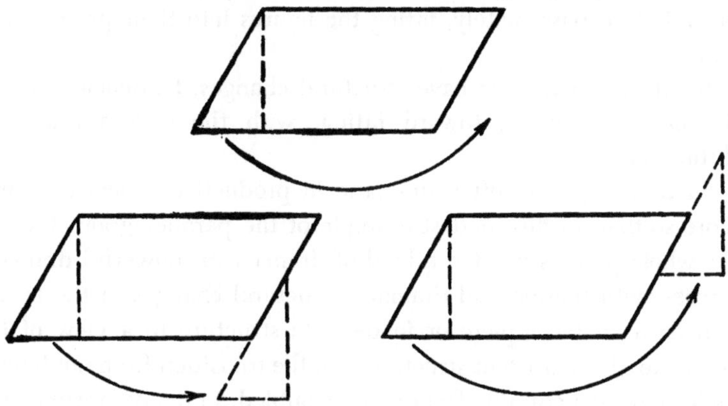

or even in this?

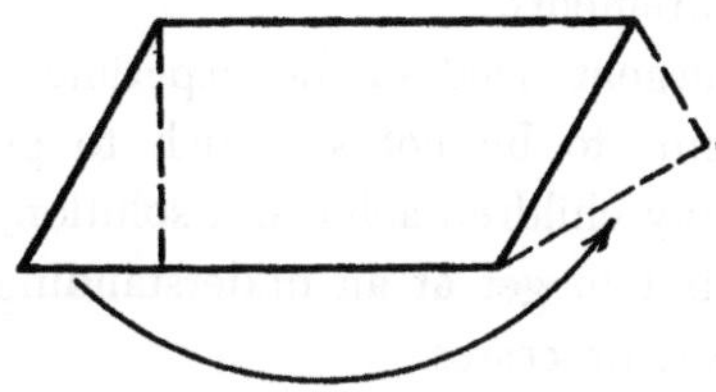

FIG. 34

Some are flabbergasted; some laugh; some passionately interfere and put it in its proper position.

It is also interesting to observe the behavior of children (even of very young children) in the following situations. Four solid figures of this kind are given:[16]

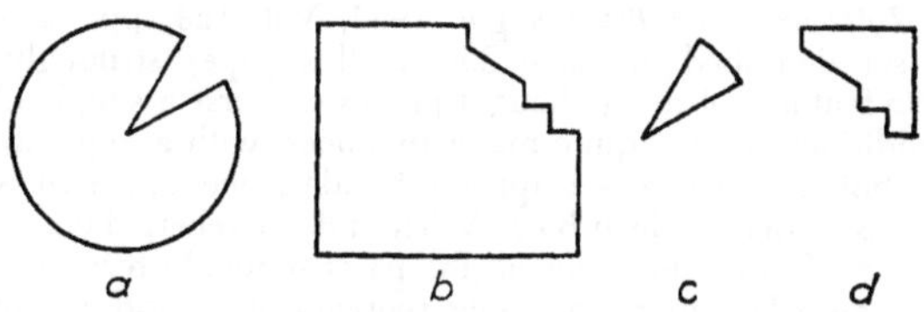

FIG. 35

[16] Cf. M. Wertheimer, "Zum Problem der Schwelle," *Bericht über den VIII Internationalen Kongress für Psychologie*, Groningen, 1926.

Children often show a strong trend to bring them together properly, to fit *c* into *a*, *d* into *b*. If the grownup tries to do it the other way, insists on placing *d* with *a*, and *c* with *b*, or puts *c* with *a*, and *d* with *b* but improperly, children are often not only puzzled, or amused, but interfere passionately, fitting the figures into their proper positions.[17]

In all these cases we have structural changes, tendencies toward the better structure, toward fitting, with the disappearance of disturbances.

Such changes are often dramatic in productive processes, much more so than in this modest example of the parallelogram. Indeed, the whole process is often a kind of drama with powerful dramatic forces—with tension and dramatic structural changes in the transition from an incomplete or inadequate structure to a view of the complete, the consistent structure,[18] in the transition from not having understood structurally, from being troubled, to really grasping and realizing the requirements.

36. The most urgent need in the experimental investigation of the problems seems to be not so much to get the quantitative answer, "How many children achieve a solution, how many fail, at what age?" etc., but to get at an understanding of what happens in good and in bad processes.

[17] It is easy to blind oneself to the real problems by simple recourse to "familiarity" of the complete figures (Cf. ¶38). Familiarity often works in the same direction as a good gestalt but a real decision is found in cases in which the structurally good figure is the less familiar and the structurally less good the more familiar. This way of deciding the issue can be generally applied to structures. Cf. W. Krolik, "Über Erfahrungswirkungen beim Bewegungssehen," *Psychologische Forschung*, 1934, Vol. 20, pp. 47-101; M. B. Hubbell, "Configurational Properties Considered 'Good' by Naive Subjects," *American Journal of Psychology*, 1940, Vol. 53, pp. 46-69.

[18] Cf. M. Wertheimer, "Zu dem Problem der Unterscheidung von Einzelinhalt und Teil," *Zeitschrift für Psychologie*, 1933, Vol. 129, pp. 353-357. The experimental set described on page 356 of this paper strikingly demonstrates characteristic features of many thought processes. First a simple clear dot-figure is shown; additions appear, quite reasonable ones, with a gap calling for proper completion; but now a new set appears besides, one that strikes the observer as nonsense, as senseless, disturbing. What sudden relief, then, what a change, when with some further additions *all* the parts suddenly form a consistent clear whole, in a new orientation, in strong reorganization and recentering, all fitting the structural requirements. Often one observes signs of strong tension in the subjects, surprise, uncertainty and, in the end, sudden relief. Afterward, when subjects describe the situation they refer to the striking structural changes in most vivid terms.

A physicist studying crystallization may try to find out in how many cases he finds pure crystals and in how many he does not—there are crippled crystals some corners of which are jagged, there are impure crystals, there are Siamese twin crystals improperly grown together, there are even crystals shaped by artificial polishing into perfect forms entirely incongruous with their nature. All such cases are of primary interest to the physicist, not as problems of statistics but for what they reveal of the inner nature of genuine crystallization.

It is also important to find out what are the conditions under which pure crystallization may take place, what conditions favor it, what factors endanger it.

And so in psychology.

IV

37. An easier way out? The role of past experience?

A wise friend, whom I told of the scissors solution (p. 48), exclaimed: "The child is a genius!" But many psychologists would say, "What of it? Obviously this is nothing but past experience. Why trouble with complicated and difficult explanations? Isn't it much easier, and more in line with many other processes in psychology, to view what these children did simply in terms of recall of past experience? By chance, or by some association mechanisms, the child recalled what she had in the past experienced with scissors. That other children failed to solve the problem is due to the fact that they did not recall, or that they had not had sufficient experience with scissors. They had not learned the connection, the association which would have helped, or they did not recall it. All that is essential in such processes is recall on the basis of learned connections. It is memory and recall that constitute such a process."

There are certainly cases in which such a thing as the use of scissors comes into play by chance or by external recall. There are cases, even in good processes, in which such suggestions from memory are tried and used, or dismissed as of no help. Unquestionably beside the present experience, whatever it may mean, much previous experience is needed in order to make such processes possible or probable.

But—is it adequate to discuss such things only in theoretical generalities? In our case, for example, the assertion is made that recall of the scissors and of associations connected with them is decisive.

Suppose a child trying to solve this problem does not think of scissors. There is a lack of this content, and of its associations. Why not take the theoretical bull by the horns?[19] Let us give what is needed and see what happens. If the essential thing is to recall the scissors, we may furnish them directly without troubling the memory to recall them. Or we may introduce stimuli eliciting the recall.

I put the scissors on the table at the beginning of the experiment or even have the child cut some paper. Sometimes it helps (e.g., if I bring the scissors after some deliberation by the child, after some remarks showing that the structural requirement is felt).

But there are cases in which it does not help. The child looks at the scissors, then looks again at the figure. Seeing the two he is clearly troubled but nothing happens.

I increase the "help." "Don't you want to take the scissors and cut the figure?" The reaction is sometimes a blank stare; the child obviously does not know what I want.

Sometimes children may obediently start to cut the figure in some way or other:

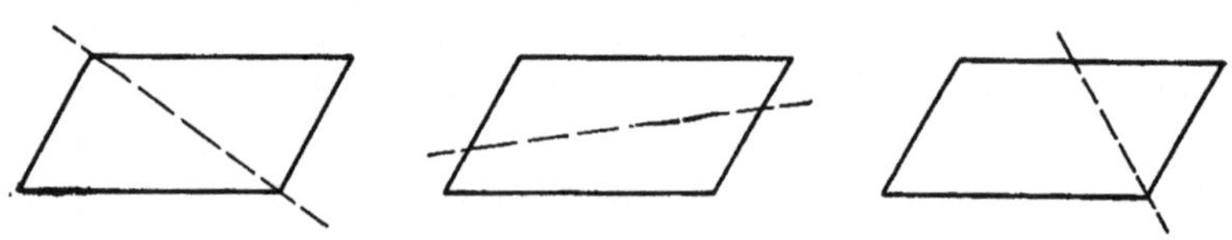

FIG. 36

Occasionally it happens that a child then puts the two parts together differently, getting another parallelogram. . . .

[19] Cf. N: R. F. Maier, "Reasoning in Humans," "The Solution of a Problem and Its Appearance in Consciousness," *Journal of Comparative Psychology*, 1931, Vol. 12, pp. 181-194.

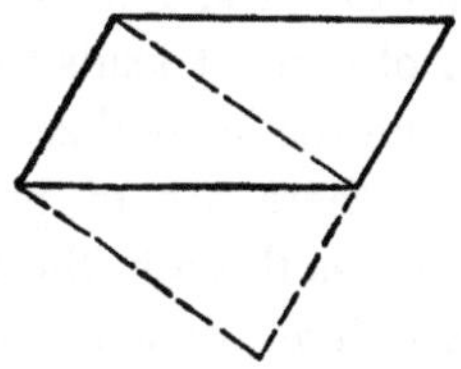

Fig. 37

When does giving the scissors help? When does it fail to help? We see that furnishing the scissors, and its common use, does not in itself help; it may result in entirely blind and foolish actions. Stating it briefly, it does seem to help if the structural requirement is already felt or if it is elicited by the scissors;[20] it does not easily help if it occurs without realization of what is required structurally, if the subject does not see the scissors in their function, in their role, in their context, with regard to the structural needs of the situation, with regard to what the situation itself requires—then he has them only as an added thing in an and-sum of contents. In fact, in some of the observed positive processes, there were remarks first that showed some realization of the structural need, leading then to recall or trial, which is quite different from blind recall of past experiences.

Moreover, not only does recall not seem to be blind, the real issue is: *what* was learned in this past experience in the positive cases? Some special and foolishly generalized movements with the scissors associated with some results of the cutting? Or the inner relation between the manner of cutting and the results? There is a ρ-relation between operation and effect, a clear fitting of the effect with the operation. This makes possible the sensible application to a new case.

Another similar explanation: the decisive thing is whether the child recalls his experiences with mosaic games which involved putting figures together and separating them into parts.

Procedure: just before giving him the problem, I have the child play a mosaic game with shapes more or less similar to those of the problem, containing within a variety of arrangements one that

[20] See N. R. F. Maier, *op. cit.*

is even partly identical with the task. This proves somewhat helpful. Nevertheless, in a number of cases it leads to no solution.

I do not know whether the reader realizes the infinite number of theoretically possible ways there are of putting things together. Even with two such triangles as those in the figure, there is a great multiplicity of possibilities, of which only few types regularly occur with children.

FIG. 38

There is a wide field here for experimentation. The observed facts do not seem to go in the direction of some, of just any arbitrary factual connections; they soon go in the direction of fitting, closing, getting a good, complete figure.

Even if a positive procedure could be explained by the working together of experienced connections, on the one hand, and of the goal-idea of the rectangle on the other, it seems to be in our case generally not just past experience but the nature and the structural fitting of past experience that has to be taken into account.

Introducing "helps" provides a technical tool for the experimenter to get an understanding of what happens. Sometimes it is more helpful to give other tasks which, in piecemeal details, may be even more complex, unfamiliar, but which have a more transparent, a clearer structure, as for example some of our A-B pairs. It sometimes happens in such cases that the subject sees the light, returns to the original problem and finds the solution. On the other hand, it is possible that he remains blind in spite of "help" which actually gives him just what he needs.[21]

The results of such experiments seem to show that it is essential that the help be seen *in* its functional value, in its place, role, and function within the requirements of the situation.

Now it becomes understandable why one can sometimes give one, two, or even all three auxiliary lines as help without their actually proving helpful. A child who does not see them in their role, in their function, may receive them as added complications, ununderstandable additions. The situation may as a result become more puzzling than before. The lines themselves need not necessarily throw light on the problem.

And was not the lesson the teacher taught at the beginning of this chapter, the extreme of such procedure? He gave *all* the items

[21] Cf. N. R. F. Maier, *op. cit.*

needed, explicitly; he drilled the pupils, achieving knowledge in routine cases but for the most part no real understanding, no ability to deal with varied situations.

It is not enough to try to substitute for the sensible process a sequence of drilled connections, even if this results in the ability to repeat, to perform what was taught in routine cases. For it would then be necessary to provide additional drill for variations in the situation, for A-B cases. It might be necessary repeatedly to furnish added mechanisms for new types of A-cases. To show that a sensible process can be replaced by a number of associations does not prove anything so long as the applications to variations in A-B cases are not taken into account and provided for. Such an enterprise is like trying to simulate the curve of a thrown ball by having it move across the open ends of a row of parallel open pipes with appropriate air pressures. By varying the pressures one could produce other curves corresponding to various angles of throwing and various weights of the ball. Or, again, it is like putting a calculating machine to work to deliver exact solutions of mathematical problems—forgetting or not realizing the piling up of devices that would be needed in the machine in order to make it properly perform some specific A-variations as well. It may be very efficient for solving routine problems, but it cannot adapt itself to new A-variations. Moreover it does not know which operations to apply; this you must tell the machine by setting the problem, by pushing the button for addition, subtraction, etc.

In short, the role of past experience is of high importance, but what matters is *what* one has gained from experience—blind, ununderstood connections, or insight into structural inner relatedness. What matters is how and what one recalls, how one applies what is recalled, whether blindly, in a piecemeal way, or in accordance with the structural requirements of the situation.

In addition to the specific structural experiences which we have in facing a problem—experiences which refer to structural perception, to changes in structural perception, to observing the results of trials, etc.—there are many general features in our world which generally play an enormous role in our dealing with objects, and which do so specifically in the concrete steps required for the solution of this concrete geometrical problem. They are so obvious that most of us do not think of them explicitly. Indeed, the reader may be shocked even to see mentioned, for example,

that in shifting the triangle from the left to the right no change occurs in the size or the form of the triangle;
that in doing so no changes occur elsewhere in the figure, no contraction or expansion in other parts;
that objects like the parallelogram, etc., have their constancy, are not changed in size by drawing lines in them;
that stated equality of some separate lines or angles secures equality of whole figures in distant places;
that cutting into parts and rearranging them in an actual operation does not change the area;
even that mere thinking operations—stating equalities, etc.—in no sense change the material;
and so on . . .

Most of the foregoing statements seem trivial and so obvious to us that they look like hidden axioms, true by necessity. They are not. If taken in connection with real happenings, they are by no means "necessary" facts. Worlds are possible in which they do not hold. Even in our own world modern science has shown them to be in many respects too simple assumptions; and in some regions of daily experience they are not true factually.

But leaving aside the question of factual truth, are they of the character of just any connections, of associations in the exact meaning of the term, such as may connect nonsense syllables? No. They have the character of simple expectations on the basis of structural context, negating sheer arbitrary, blind connections. To make the point concrete, as long as no other factors come into play, it is structurally simplest, most reasonable, not to expect such changes as, e.g., a queer contraction of, say, 7 per cent somewhere in the right part of the parallelogram when it is cut at the left.

If we ask whether these features are learned, are acquired in accordance with the traditional associationist conception of past experience, this seems improbable indeed in the light of experiments concerning gestalt laws, etc. They center in laws of organization and reasonable structure; they are an outcome of the structural way of working of our mind and brain rather than of blind associations.[22]

Thus the mentioned hidden axioms appear by no means as simple products of sheer association which may connect any items, irrespective of their inner relatedness, of their structural features, of reasonable requirements.

Still other, entirely different factors of experience play a role in such thought processes, and an important role. The attitudes one has developed in dealing with problem-situations—having had the experience of achievement or only of failure, the attitude of looking for the objective structural requirements of a situation, feeling its needs, not proceeding willfully but as the situation demands, facing the issue freely, going ahead with confidence and courage—all these are characteristics of real behavior, growing or withering in the experience of life.

Thus problems of personality and personality structure, structural features of the interaction between the individual and his field are basically involved. In connection with the latter we have also to realize the structure of the social situation, the social atmosphere one is in, the "philosophy of life" developed in the behavior of the child or person in his surroundings; the attitude toward objects and problem-situations eminently depends upon these factors. So also the social atmosphere in the schoolroom is sometimes of considerable importance for the development of genuine thinking. In the solution of this kind of problem it is more helpful at times to create the right mood than it is to force on the subject certain operations or drill.

Interested in clarifying certain fundamental features, we limited the scope of our discussion. We were able to do so since we were dealing with a relatively closed sub-whole. But if we want really to understand how the performance comes about (or does not), a much broader field must be confronted. The question then is one of the organization of the whole field in which the actual happening is only a part[23]—the personal, the social, the historical field. As

[22] M. Wertheimer, "Untersuchungen zur Lehre von der Gestalt," II, *Psychologische Forschung*, 1923, Vol. IV, pp. 336 and 349; see also W. D. Ellis, *op. cit.*, selection 5.

[23] Cf. M. Wertheimer, "Über das Denken der Naturvölker, Zahlen und Zahlgebilde," *Zeitschrift für Psychologie*, 1912, Vol. 60, pp. 321-378; reprinted in

to the last, our present generation stands upon the shoulders of previous thinkers in long historic developments. These are large tasks. I regret that I cannot deal with them explicitly here. In all these realms structural issues are involved no less, I think, than in our little examples. Some work has been done in these directions; much more is needed.

There are still psychologists who, in a basic misunderstanding, think that gestalt theory tends to underestimate the role of past experience. Gestalt theory tries to differentiate between and-summative aggregates, on the one hand, and gestalten, structures, on the other, both in sub-wholes and in the total field, and to develop appropriate scientific tools for investigating the latter. It opposes the dogmatic application to all cases of what is adequate only for piecemeal aggregates. The question is whether an approach in piecemeal terms, through blind connections, is or is not adequate to interpret actual thought processes and the role of past experience as well. Past experience has to be considered thoroughly, but it is ambiguous in itself; so long as it is taken in piecemeal, blind terms it is not the magic key to solve all problems.

38. Résumé with regard to the question left unanswered in ¶10:

Let us return now to the question we had left unanswered in the end of part I, ¶10, the problem of the A-B responses. Our discussions in the previous paragraphs indicate a direct answer.

The teacher has shown the procedure: he has taught the class to draw the three auxiliary lines (p. 15). *If* the pupils have really grasped the issue, then for them the three lines are not just "this line, and that line and the other line" or, as the teacher had said, "a vertical line drawn from the upper left corner, another from the upper right corner, and a prolongation of the horizontal line from the lower right corner." They are not an and-sum of items, blindly connected with the solution. If that were what the pupils had got from the lesson, they would be lost in the critical A-B tasks, they

Wertheimer, *Drei Abhandlungen zur Gestalt Theorie*, (Erlangen, 1925); see also W. D. Ellis, op. cit., selection 22; H. Schulte, "Versuch einer Theorie der paranoischen Eigenbeziehung und Wahnbildung," *Psychologische Forschung*, 1924, Vol. 5, pp. 1-23; K. Lewin, *A Dynamic Theory of Personality* (McGraw-Hill, 1935); E. Levy, "Some Aspects of the Schizophrenic Formal Disturbance of Thought," *Psychiatry*, 1943, Vol. 6, pp. 55-69.

would have no basis on which to deal sensibly with the new problems.

But if they have grasped the problem—and this is what grasping means—then they see the lines *in their structural role and function,* in their meaning within the sensible context. They *see how* these lines, just these in this situation, bring about the solution in the *inner relatedness,* the structural relation ρ of these operations *to* reaching the goal. The operations are viewed "from above," from the vantage point of the inner structure of the whole procedure, as they function within the context and fit its requirements. This then gives a basis for sensible dealing with the A-B problems.

Two things that are related are decisive here: the structural meaning of the parts and the character of their inner relatedness with reaching the goal, actually, transparently. The first has been clarified in many aspects; the issue in the second is a ρ-relation between the operations and reaching the goal.

A first approach may view our problem in terms of what the lesson learned has provided as to structural transfer in changed situations. Suppose we speak of drawing the three lines as the "learned means to the end." In the figure used in teaching, situation s_1, means m_1—drawing the three lines—leads to the goal, g. This is learned:

$$s_1, m_1, g.$$

What has been provided as a basis, as help for finding the appropriate m_2 in a situation s_2, the appropriate m_3 in s_3, etc.? What has been provided as to structural transfer of the m in changed situations?

Obviously we have to discriminate. m_1, objectively the same, may yet function in different ways: if the three operations have been learned only as an and-sum, without grasping or focusing on the inner, structural relation between just this kind of m in this situation, and successfully reaching the goal, then we have a set of operations that may be repeated, applied correctly to routine variations in some sort of structural transposition, or by blindly following the abstract wording of the teacher according to which the length of the lines is variable. The problem may be solved as long as the variations in s allow applications of just those lines. But if they do not fit the new situation, then there is no basis in what has been learned for dealing with it. In other words, if the meaning of the three operations is centered only on the wording of the teacher (two verticals from the upper corner, prolongation of the horizontal to the right), then the length of the lines and the distance between them are variable and adjustable to a routine form s_2; *but* for situations in which these three general helps are not applicable, which require that *they* be changed, nothing is provided.

Contrariwise, if the procedure is structurally understood, solving is centered in quite a different way and the resulting structural transfer is something fundamentally different. If the center of the procedure is the structural grasp—getting rid of a disturbance in the figure by the compensating shift—then the new situation too may be viewed in terms of looking for the disturbance, for the gap, and for what is required here in order to deal with them. Accordingly, the number, the length, the place of the auxiliary lines are variable, and they may change according to the features of the new situation.[24]

As in the genuine processes (pp. 48-49) the steps came about on the basis of the structural trouble, the structural requirements, here the responses to these changed situations come about sensibly, with the help of what has been understood in the learning situation.

There are characteristic cases in which a subject has not really grasped in the learning situation. He has succeeded in routine variations, applying what the teacher had shown; but confronted with new tasks, he is lost. Spontaneously he goes back to the taught lesson, ponders over it, then suddenly cries: "I've got it!" and having understood the s_1, m_1, g structurally, he goes at this new task and he solves it easily. Subjects often describe in the most vivid terms what happened to them in the transition from copying what the teacher had taught to "seeing the light"—how drawing the three lines in the teacher's example suddenly became transparently clear,

[24] E.g., in some cases (see example p. 48) m_2 easily becomes two lines instead of three. For the case on page 17 a different position is sought which allows the clear interchange of the disturbances. In the case on pp. 18-19 a hint is provided to search for interchangeable parts, out of which the idea of bisecting the slanting line may arise.

meaningful, in the realization of the inner structure, the inner requirements of the process. "And then it was easy to deal with the new problems."

In short, this is our formula: in real A-reactions behavior is determined by the requirements of the given situation, in B-reactions by some external details. In A-reactions the subject deals with the new situation on its merits, having understood the lesson situation on its merits.

The problem of transposability is of some importance and, although I think that the reader who has followed through sees the decisive point, I may add that the problem is by no means settled by stating the general formula. A number of problems arise for the scientist: here is a broad field in which experimental investigation seeks clarification, the conditions and laws governing the extent and character of variability that result from different learning situations. In order to understand the problem, it is necessary to study it *in contrast* with cases in which the learning gives no material aid for dealing sensibly with changed situations, in which even the greatest genius could find no basis for sensibly transposing the well-known and much used "rote" learning situation.

On the other hand, there is the possibility that the subject grasps the inner structure of the situation, which will then help him when he is dealing with variations of the original problem. Let us consider an entirely crude case of s_1, m_1, g in which no such possibility is given. Suppose, instead of drawing the three lines which bring about the transformation of the parallelogram into a rectangle of equal size, a subject is shown a parallelogram on a screen; furthermore, there are several buttons and the subject learns that by pushing a red button, a blue button, and a green one, the parallelogram disappears and a chocolate drops out—or a rectangle appears on the screen. He may learn that. But if afterward you give him another figure—of the A- or B-type—naturally he will be at a loss. He may try to push the same buttons but with no result. He may try other buttons in blind trial and error, he may even by chance hit on the right ones; but he will again be lost when another figure is presented, for there is no possibility of discovering a sensible inner relation in the s_1, m_1, g. The connections are arbitrary, or hidden, and no basis for sensible variations can result.

Many theorists are blind to the problem, to the difference between such cases and cases which are solved sensibly. They have an easy ready-made way of avoiding the issue; they notice, and notice correctly, that, in the former, help from past experience is precluded, not in the latter; and they conclude—but the conclusion is blind—that the difference in the former responses is explained simply by the working of previous associations, all basically of the same nature as in rote learning. Sensible learning and applying is to them nothing but the work of a pile of previous associations. I hope the reader sees after all our discussions that this would be dismissing the problem far too easily: even if all the factors involved were due to past experience, our problem would remain. The crucial question is not *whether* past experience, but *what* kind of past experience plays a role—blind connections or structural grasp with resulting sensible transposing; also *how* material gains from past experience come in, whether by external recall or on the basis of structural requirements, of material functional fitness. Recourse to past experience does not therefore settle the problem; on the contrary, the same problem repeats itself with reference to past experience.

To study the use of what one has gained in past experience is of profound interest; but for our problem, in a first approach, it is not decisive whether the material used stems from present or from past experience. What is important is its nature, whether a reasonable structure is grasped, and how it is brought in. Even if everything, including grasping itself, were due basically to repetition of past experience—a hope some psychologists entertain, but which I regard as misleading, or at any rate unproved—or if we were to apply the concept of drill learning even to reasonable structures, it would still be important to face and to study the described difference, as it is decisive for the possibility of structurally reasonable processes. In everyday language, to have gained experience means for the most part something quite different from piling up external connections of the character of rote learning in the last example; more sensible gains, more sensible processes are meant.

We can summarize the A-B questions relating to the parallelogram as follows: for the role which a given s_1, m_1, g plays in facing new situations, the decisive point is *what* is learned from the taught example as well as from other past experience. Whether such experience established blind connections or insight into the structural issue is the alternative which determines what has been provided for dealing sensibly with A-B variations. To this I may add that the role of the particulars in s_1, m_1, g, may be of greater or of less importance; the extreme case is reached when the gain for new tasks is mainly the fine experience of being able to go ahead by looking for the requirements of the ***situation at issue*** and trying to deal with them straight.

39. If in such processes one looks for the operations of traditional logic, one can find quite a number of them. One can even describe the process in a succession of propositions. But the aggregate of these propositions does not represent what really happened in the process. Much has disappeared. The dynamics, the life is gone.

Traditional logic is not so much concerned with the process of *finding* a solution. It focuses rather on the question of correctness of each step in the proof. Here and there in the history of traditional logic there have been hints of how to proceed to find a solution. Characteristically these attempts have been of the following nature: "Look for any general propositions you know, the content of which refers to some of the items in question; gather these propositions together; look for pairs among them which, containing a common concept (middle term), permit of a syllogism," and so forth. (Cf. the example in Chap. II, p. 83, which, though foolish, is much in line with such a procedure.)

We shall return to the issue of the proof; we shall then see briefly that the sensible proof itself involves structural factors. But for the moment let us examine some characteristic aspects of the attitude of traditional logic by considering the following remark of a logician: "The whole thing comes down to the use of the principle of commutability, $a+b=b+a$, just as $2+5$ equals $5+2$; both give the result 7." (An empiricist might have formulated it in much the same way.)

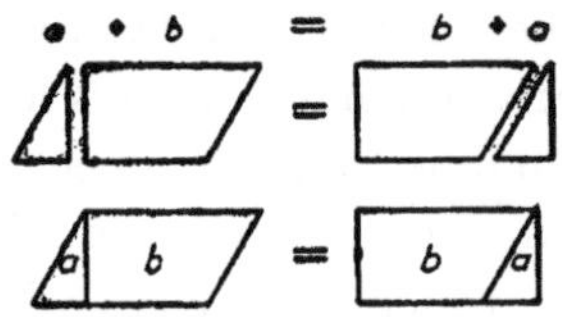

FIG. 39

Think it over yourself, reader. Compare this assertion, in the spirit of traditional logic, with a genuine finding process. Maybe you will agree; maybe not. If you see differences, are these unimportant, peripheral? Or do they involve factors that are decisive for just this problem of productive thinking? If you are a logician, accustomed to the ways of traditional logic, very sure in your definitions of what logic is and what thinking is, you may react violently against some of the remarks that follow. Please do not make use of the habitual escapes and dismiss the matter; try to do justice to the points I am going to mention. Do not misunderstand: this is in no sense an attack on the correctness of traditional logic. It is an invitation to realize certain problems and to see the teachings of traditional logic in their proper place.

The principle $a+b=b+a$ is somehow involved in the process of finding the area of the parallelogram, but in a way quite different from its meaning in traditional logic. And just this difference seems essential for making possible the genuine productive process.

1) First let us mention briefly that the *a* and the *b* in the figure above are *not* there from the start. This division of the parallelogram has to be reached in the process of solving the problem! And it is quite important that just this triangle *a*, just this kind of division is found, created, whereas for the formula that is no issue, the *a* and *b* are there, ready-made from the beginning.

2) A second point: while $a+b=b+a$, *a* is supposed to remain unaffected by the change of position, in actual thinking about the area of the parallelogram the triangle *a* changes its meaning as it is shifted. The *a* at the left side is found, created in the desire to get rid of the disturbance. The *a* at the right side of the equation is the

part needed for completing the gap. Only with regard to bare equality of size do we have an equation; the equality of size is important but the transition from the left to the right is a transition from one thing to quite another thing: $a+b$ is not the same as $b+a$ in regard to form, and quite essentially not the same for the process itself.

Even if we abstract from the actual processes, the formula $a+b=b+a$, taken exactly, is not the equivalent of the equation expressed in the diagram (see diagram below). It would be really adequate only if the two parts a and b had nothing to do with each other, were just two pieces, the mutual positions of which were of no importance whatever. But the form is important—or we would have no parallelogram and no rectangle either.

Examination of the parts in the figure clearly shows that there are striking differences between the left and the right of the figure. This holds not only for the whole-features—the parallelogram versus the rectangle—but also for the single parts of the figures. If the reader will study and compare the meaning of the lines,

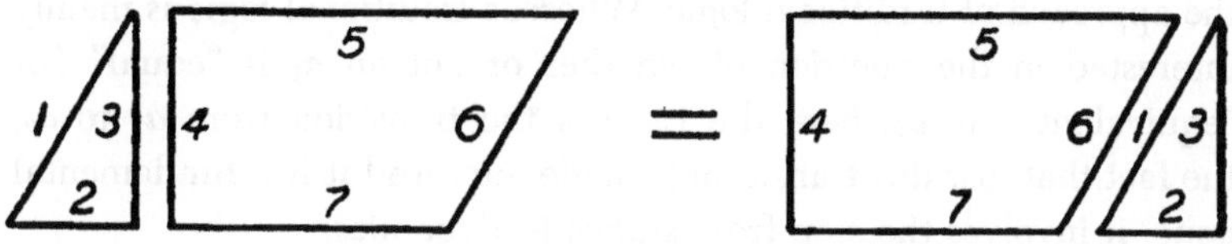

Fig. 40

he may be surprised to realize how strikingly different the situation is with regard to them at the left and the right. I shall mention only a few points. The lines 1 and 6 at the left are boundaries; at the right they become identical, are no longer two lines, as they disappear from view in the completion of the rectangle. At the left 1, 5, 6, 2—7 form the figure, with the added, introduced line 3—4, whereas at the right side, 4, 5, 3, 7—2 form the figure and 6—1 is a line that disappears from the picture. The equation ignores the fact that together these lines form the periphery which, after all, is important for the figures the size of which must be determined.

Similarly with the angles: their meaning, their function in the figures is basically different on the two sides; angles which play an essential role at the left disappear at the right, etc.

If one makes the exact analysis of all such factors, one arrives at quite a huge number of structural differences. Viewed from below, piecemeal, such structural differences would look extremely complex. It would be very hard indeed, in all probability impossible, to get the clear process if one were to start with a mere sum of such detailed features. But there is no disturbing complexity if the approach is "from above," from the whole-qualities of the figures and the functional meaning of the lines, etc., in them.

3) Basic in the productive process is the *change* that results when $a+b$ becomes $b+a$. We have not in the figures, as in the equation, simply a relation of mutual equality of two things, but a change with a *direction*,

$$a+b \rightarrow b+a$$

and a *required* change. It is a change from one thing to something quite significantly different. We have not just an equation, we have a *transition*. The problem of validity, though important, is fundamentally blind to this directedness. Here is a basic difference from the approach of traditional logic. Whereas traditional logic is mainly interested in the question of whether or not an a_1 is "equal" (or "equivalent") to a_2, here the issue is the transition from a_1 to a_2, the fact that just this transition is made, etc. And it is a fundamental issue; it involves the step from statics to dynamics.

Do not such transitions involve the alternatives, "logical" or "illogical," sensible or blind and arbitrary procedure? Should these not be a matter of logic?

Such a "transition" often involves "structural reorganization." I may remark here that this term, important in gestalt theory, is often misunderstood, so that its meaning is cheapened. Some years ago a psychologist illustrated its meaning by a series of nonsense syllables to be learned first in one sequence, then in another. No such arbitrary procedure is meant here, but reorganization as sensibly required by the structure of the situation. The vectors for and in this change grow out of the functional requirements of the structure of the situation.

And I may mention that in such genuine cases as this, the transition seems inadequately characterized by stating that it is a transition to a more familiar figure; the transition here is to a form in which the matter becomes structurally clear. The size of the area in terms of unit squares becomes transparently clear in the form of the rectangle.

4) It may be noted that the equation $a+b=b+a$ does play an important role in the process with regard to the question of mere size also. The principle that such operations do not affect size is an outstanding case of structural simplicity in contrast to the possibility that such operations could be accompanied by changes in size (see p. 63). This does not mean that it is true by necessity. Nature is not obliged to show such simple properties. What is comfortably true with regard to the sum—we are dealing here with the *size* of the area, which is an and-summative issue—is not simply true generally, is not true for issues that are not of an and-summative nature.

The order *ba* versus the order *ab*, though irrelevant to the and-summative question of mere size, is very relevant in other aspects of this process. In fact, the order of the succession often does much more to the thing, to the nature of the parts, to the dynamics involved, than in our case. In our example, after all, one again gets a closed figure in the change. Contrast the two ways in which *ab* is changed into *ba* in these simple instances:

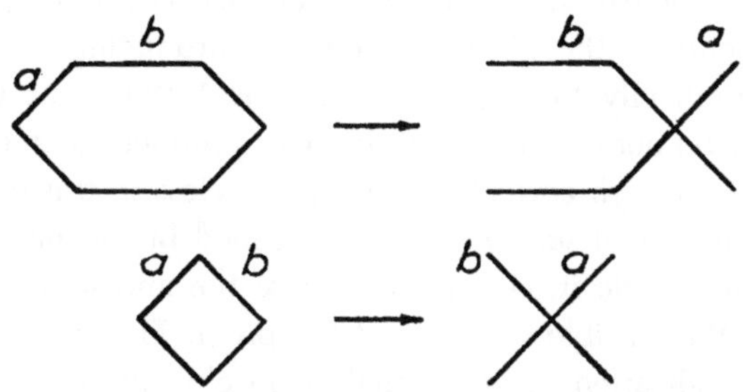

FIG. 41

And to assume the principle of commutability, e.g., in melodies, would be quite blind, quite senseless. The same is true in many other cases. Great problems are involved here, fundamental questions in logic. Some of the problems, such as those illustrated above in the hexagon and the diamond, are partially studied in the modern theory of relational networks, etc.; but deeper problems arise in regard to whole-qualities and whole-dynamics.

Many still view the principle of commutability as a general, basic principle in logic, generally presupposing the summative, atomistic character of facts, of propositions, etc., to such an extent that the doctrine could arise that logic fundamentally deals with "tautologies." In the light of our discussions such a view seems fundamentally blind to real problems in thinking.

The principle of communtability certainly does not hold true generally for the items in a real thought process. If one were to put the items, operations, or steps in a genuine thought process into some mixed-up order and state equality in terms of the principle of commutability, it would be quite a blind statement. The items of such a process are not an and-sum of pieces.

5) For the logician the principle of commutability is one of the statements among the propositions which make up the proof. Now it seems necessary to remark that the proof itself has its structure. If a subject is blind to the structure of the proof, sad things happen. Often a pupil, facing the series of statements in the proof in his textbook, is puzzled, worried, at a loss. He reads the statements, checks them in the diagram, reads the theorems, starts a procedure of trying to fit the pieces together as in a jigsaw puzzle, to get at the sensible context. If he still does not succeed, he may memorize the statements in the given sequence; reproducing the proof, he may try frantically to recall what the next statement was in the textbook; if he does not succeed, he may produce some other statement which, though correct, is simply senseless in this context. The bright pupil, of course, does what is needed but in many textbook proofs *he* has to do it. He has to change the and-sum of the statements into the sensible structure of the proof. This involves sensible grouping, realization of functional hierarchy, of direction, of the structurally sensible place, role and function of each statement, its meaning in the whole. If a subject fails to see, for instance, that one of the statements, together with certain others, belongs to one sub-whole in the proof (say, the congruence of the triangles), and groups it wrong, he is far from seeing the light. There are cases in which a subject, trying to get some order into the statements about

lines only, then about angles, then planes, is rather proud of now having a logical order, but becomes desperate again as he recalls the task. It is of no little importance to realize whether a statement has the function of a premise or of a conclusion, itself serving as a further premise, etc.

Similar considerations hold for *finding* the proof. Sensibly finding the proof does not proceed in the manner described above (p. 70) which is so characteristic of the attitude of traditional logic. It is not just a matter of making correct statements, recalling some learned theorems, etc. The sensible discovery proceeds by realizing the requirements for the proof, by facing what is needed in a structurally sensible order.

But whereas the structure of the proof in our case of the area of the parallelogram is comparatively simple, it is not nearly so easy to find a psychologically adequate, structurally sensible procedure in other teaching matters. Here improvement, productive work, is urgently needed.[25]

40. We have discussed the factors which are essential in solving the problem, reaching the goal. But what of the goal itself? Often thinking processes are viewed as processes of solving a problem, reaching a set goal; and so far we have done so ourselves. The task of thinking consists, in many theories, of just that. But are not our problems repeated with regard to the goal itself?

Here, in our case of a modest geometric problem, the situation is generally innocent enough. The very joy in solving such a problem is at work, the joy in achievement, in using one's mental powers. To this extent thinking may be a relatively closed affair. Moreover, there are situations in which the task is sensible in a larger context. This is the case when the problem of the area is set in a context in which a farmer wants to determine the area of his field; or when

[25] For several years I have been dealing with these tasks in lectures on teaching and learning and in research with a number of fellow workers.

Dr. George Katona has dealt with some of these questions in his book *Organizing and Memorizing* (Columbia University Press, 1939) and in the following papers: "On Different Forms of Learning by Reading," *Journal of Educational Psychology*, 1942, Vol. 33, pp. 335-355; "The Role of the Order of Presentation in Learning," *American Journal of Psychology*, 1942, Vol. 55, pp. 328-353. Dr. Catherine Stern reported on her work in the teaching of arithmetic in a paper read at the 1941 meetings of the Eastern Psychological Association; this will form part of her forthcoming book.

the question arises in a broader context of geometric thinking—as, for instance, when one has grasped the area of the rectangle and is interested in determining the area of other figures, in the general problem of area-determination.

But there are situations in which it is quite foolish just to work on the task of the area of the parallelogram because the task of the area does not structurally fit the situation, because the goal itself is out of place, and the situation calls for other things. If in such a situation, the task is put or the question arises somehow, some people, blind to what is needed, start to work and stick blindly to the set goal. On the other hand, we often find quite wise reactions, the subject refusing to do the problem, because he is concerned with what really matters in the situation.[26]

I shall give a simple example. A teacher quite innocently uses some occasion for practice problems. In the last lesson he has taught the class how to get the area of a trapezoid by drawing the auxiliary lines; he has taught the formula $\frac{a+b}{2}h$. He refers now to a framed picture hanging on the wall, saying:

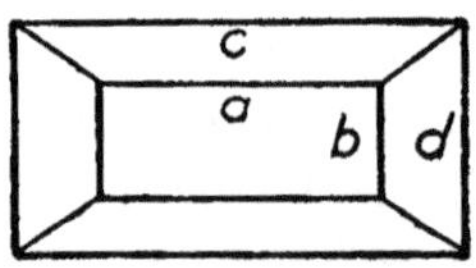

FIG. 42

"I want the area of the frame around the picture"; he labels the lines *a*, *b*, *c*, *d*; gives the dimensions, and adds: "You see, here are four trapezoids. I hope you remember how to find the areas."

Some children work busily on the task set by the teacher; tediously they determine the areas—some falling into errors, correcting them with strained attention. But some children just look amused, do nothing of the kind, multiply *c* by *d*, *a* by *b*, subtract *ab* from *cd*, saying, "So what? Why should I get the areas of these trapezoids?"

Thinking is not merely to solve set problems. The goal itself, as a part of the situation, may be structurally sensible or foolish. Just as the operations within a real thinking process function as parts

[26] Cf. the case in Chap. III, p. 111 ff.

in their place and role with regard to structural requirements, so does the goal itself, as part of the broader context. Often the thinker, in the course of trying to solve a set problem, stops, realizing that the situation requires quite different things, requires changing the very goal. To stick to set goals, to insist on reaching them, is often sheer thoughtlessness.

In life such cases are often of a serious character. Sometimes men, for instance, politicians, after trying hard to reach a certain goal and working at it a long time, suddenly realize that the goal itself, as set, was out of place, unrelated to the real requirements, to more essential goals. This in itself may be a discovery of something that was not at all realized before—namely, that the means for a sought goal would endanger, would kill a much more important goal. Thinking is not merely concerned with means; it concerns the ends themselves in their structural significance.

In our geometric tasks such questions are not so serious: these are tasks in a quiet, pure, peaceful, clear region of life, tasks in which it is possible to proceed in a transparent, crystal-clear way. This was one of the reasons why educators again and again urged the study of geometry, which allows the development of mental abilities in an atmosphere of clear, transparent consistency, and which may then help toward similarly straight attitudes in thinking about more complex and less concise matters.

This is one of the reasons why in this book we chose these simple geometric examples for discussion; it seemed better first to discuss the basic theoretical issues in structurally simpler materials.

CHAPTER II

The Problem of the Vertical Angles

HERE is an elementary question in geometry. Two straight lines intersect and form two angles, *a* and *b*. Can you prove that they are equal?

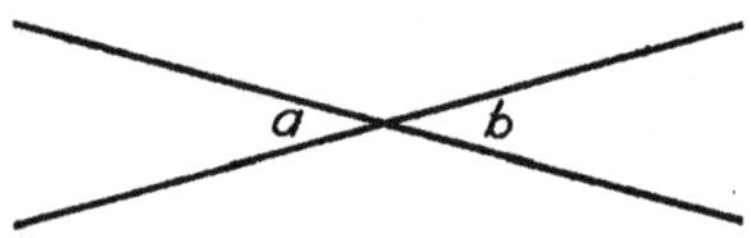

FIG. 43

Very likely you have learned this theorem in school. Perhaps you have forgotten it; all the better. Try to do it before reading what I report in this chapter. You may enjoy what follows more.

Putting this question to bright children and to grownups, one often gets reactions such as the following: "What are you asking? Isn't it obvious? Naturally the angles are equal; can't anyone see that?" And if one insists, one may get the answer: "It is clear; two straight lines first come together, then diverge in the same direction."

One of the main difficulties in dealing with the problem is that the pupil does not see—cannot see—the sense of the question. It appears artificial, senseless. Often the demand for a proof cannot be understood in such a situation; many do not know or are not able to realize the value of a proof here, the need for which came into existence in the development of theoretical mathematics.

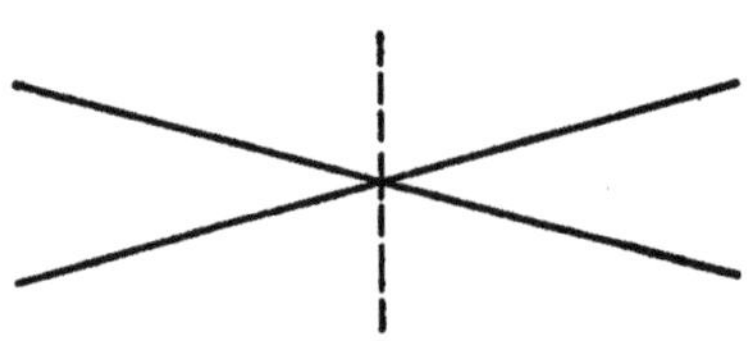

FIG. 44

Some say: "Of course you can prove it if you want to. Cut the paper vertically, turn the half page around putting one angle over the other. Hold them against the light. You will see that they coincide." If I say: "I agree, they will, but can you demonstrate here, on the figure, that the two are equal?" then for the most part subjects do not know what to do. Some fall into deep thought, which may contain little that is positive.

I shall tell first of what happens in schools.

I

The teacher demonstrates the theorem. He draws the lines, labels the angles, and proceeds as follows:

$$\begin{aligned} a + b &= 180^\circ \\ b + c &= 180^\circ \\ \hline a &= 180^\circ - b \\ c &= 180^\circ - b \\ \hline a &= c \quad \text{Q.E.D.} \end{aligned}$$

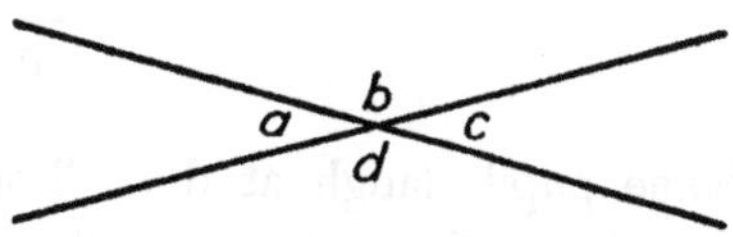

FIG. 45

One can describe the process in terms of traditional logic or of association theory. The teacher shows a number of successive operations, makes additions, writes equations, transforms them, finally derives the result. He may start with axioms or some general propositions, and apply them to the case in question. The pupils learn the proof and are then able to repeat it.

Certainly the proof can be described in terms of a number of operations and, for the problem of validity, these have to be considered. But is such an aggregation of several operations what really matters?

Some days later the teacher calls a pupil to the blackboard and asks him to demonstrate the equality of the angles. If the pupil now repeats word for word what the teacher has taught him, one is at a loss to know, "Does he repeat what he has heard blindly, like a slave, or has he grasped, has he understood?"

It sometimes happens that the pupil does not recall exactly, writes

$$a+b=180^\circ$$
$$c+d=180^\circ$$

then boldly: therefore $a=c$. Others become lost, look desperate, or get a stupid, sheepish expression. Still others may do

$$\begin{array}{l} a+b=180^\circ \\ \underline{b+c=180^\circ} \\ a=180^\circ-b \\ b=180^\circ-c \end{array}$$

and then become equally helpless.[1]

But you also find such performances as:

$$\begin{array}{c} a+d=180^\circ \\ \underline{c+d=180^\circ} \\ a=c \end{array}$$

Some pupils laugh at this: "Look! He has made two mistakes!" But the really superior pupil says, or appears to say to himself: "Why should I bother with the terms? It doesn't matter which way I work it." The teacher asks whether he cannot write the proof exactly as it was given and he writes with assurance

$$\begin{array}{c} b+c=180^\circ \\ \underline{c+d=180^\circ} \\ b=d \end{array}$$

This is certainly fresh, but obviously quite different from the changes made by the first pupil.

We see that the "number of errors" is not decisive. One error makes the response a B-case, a senseless procedure; two "errors" may or may not lead to success, the operations may or may not be sensible. Two "errors" may simply show sensible understanding. What is decisive? We shall return to this point later.

There are pupils who are seriously disturbed if the teacher so much as uses a diagram with other than the habitual labels. This is no proof of the assertion that "mind is ruled by habits through-

[1] Cf. Chap. I, p. 16 ff. Such foolish performances do not seem to be characteristic of children's behavior in general; they seem to become possible mainly on the basis of teaching by drill, etc.

out."[2] It is proof that these particular individuals stick blindly to "what they were taught." Others may be a little surprised at the change, but what they are trying to do is a different thing from slavish, senseless repetition.

Here are examples for testing A- and B-solutions.

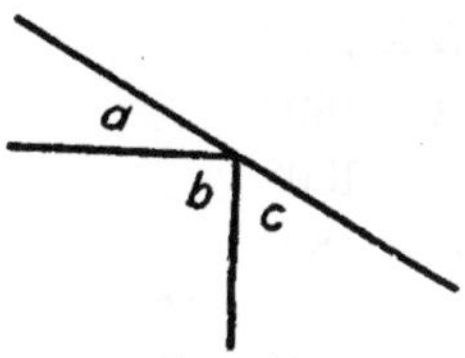

FIG. 46

1. We have a straight line; two other lines form a known angle, e.g., 90°. If one bravely applies here the steps of the learned demonstration, one shows that one has understood nothing.

This is a B-task.

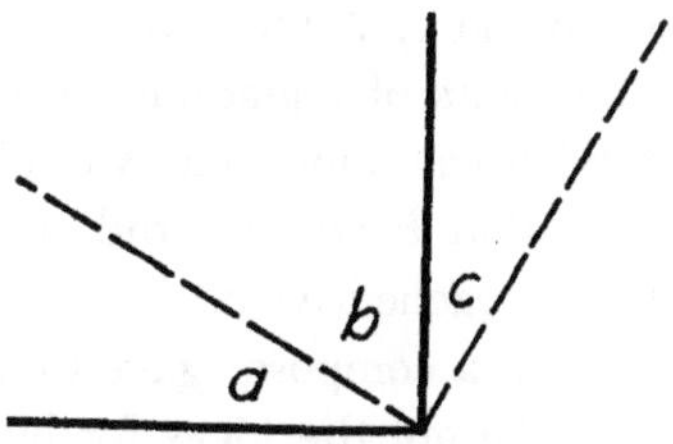

FIG. 47

2. There is a right angle. The two dotted lines, too, form a right angle. Some refuse to make an attempt, "But, teacher, we haven't learned this." But some proceed consistently, in spite of the much changed situation.

This is an A-task.

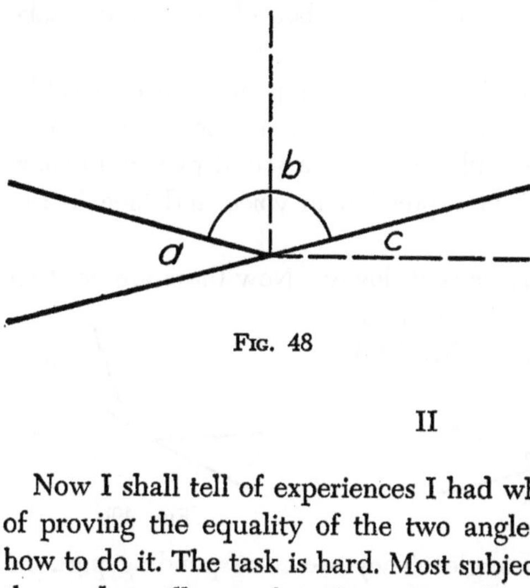

FIG. 48

3. The angle *a* is drawn and one of its lines is prolonged, giving the *b*-angle. *b* is bisected by the dotted vertical line. The fourth line is added, forming a right angle with the bisecting line. The task is to demonstrate the equality of angles *a* and *c*. The reader may discover for himself whether this is an A- or B-case.

II

Now I shall tell of experiences I had when I gave the initial task of proving the equality of the two angles, $a=c$, without showing how to do it. The task is hard. Most subjects do not succeed. I hope the reader will see why: the structural operations required are not

[2] E. L. Thorndike, *The Psychology of Algebra*, p. 458 (Macmillan, 1920).

easy to envisage (cf. p. 86 ff). I shall illustrate by three examples:

1. I shall tell first of a subject (adult) who proceeded much in accordance with some classical remarks in traditional logic. He said: "Let's see, what general propositions have I at my disposal?" After a while he became busy writing down true propositions:

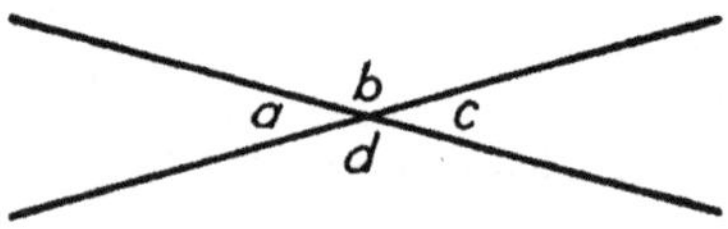

FIG. 49

$$a + b = 180°$$
$$a + d = 180°$$
$$b + c = 180°$$
$$c + d = 180°$$
$$a + b + c + d = 360°$$
$$(a + b) - (c + d) = 0$$

Then he searched for more such propositions. After a time he started making permutations, combining pairs of equations, adding them, subtracting them, seeing whether something would turn up. He finally arrived at the equation $b = d$, but did not think of stopping and proceeded until he reached $a = c$.

The procedure was similar to the answer a composer gave to a curious visitor who wished to know how he got the ideas for his melodies, how he found them. The composer, bored by the visitor, said: "Oh, that's simple; I take a number of notes and make permutations."

2. Here is a fine example of a process going ahead sensibly. The subject, fortunately, was thinking out loud (sometimes mumbling). I am sorry that I could not make a sound picture to show the illuminating changes of expression in voice and face in the course of his work.

Looking at the drawing he said slowly: "Now these are not two separate angles whose position is arbitrary with reference to each other." Asked what he meant, he drew:

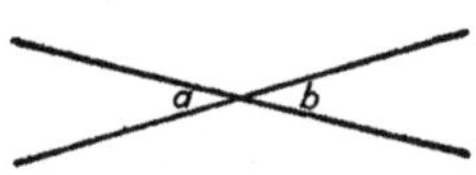

FIG. 49a

FIG. 49b

"They are not like those angles. They are corresponding parts in the figure. The straight lines go clear through. This straightness of the lines must have something to do with the equality of the angles! . . . Straightness in terms of angles means—180° . . ." Then he drew:

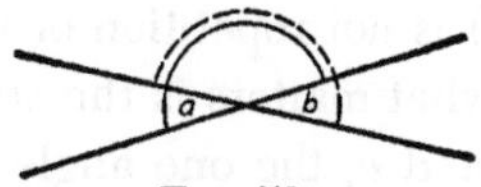

Fig. 50

and he said: "*What a is as part of its 180°, b is as part of its 180°!* The remainder in both cases is the angle above; identical in both!" He labeled it *c*, and wrote the two equations:

$$a + c = 180°$$
$$b + c = 180°$$

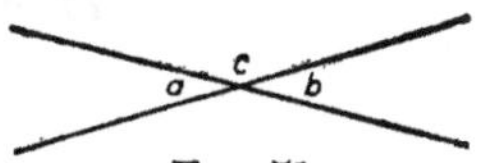

Fig. 51

He continued: "What a is in $a + c$, b is in $b + c$, and wrote:

$$a = 180° - c$$
$$b = 180° - c$$

"Therefore," he concluded, "$a = b$."

3. In another procedure which was somewhat similar at the beginning, the last steps were different. The subject saw the requirement to understand a and, similarly, b as part of 180°. But he did not at first see the requirement with regard to the remainder. He proceeded as follows: "I have to use a as part of 180°; I have to use b as part of 180°." He drew

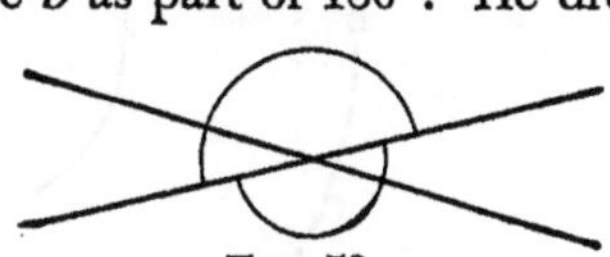

Fig. 52a

Then he hesitated, saying, "There is another possibility of pairing." Beamingly, he changed the figure into:

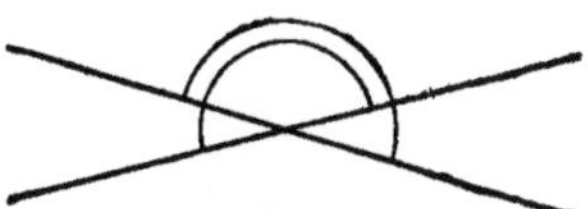

52b

III

A sensible process such as was described in our last two examples involves operations of grouping, of grasping structure, equality, symmetry, of "playing the same role," of having the same function in the group, of realizing relations, more particularly the ρ-relation in which is realized the inner relatedness with the given structure of the needed grouping.

Probably the reader has already seen what is essential in the A- and B-cases and reactions. What matters in the A- and B-reactions (see Figs. 46-48, p. 82) is not repetition of items, copying a learned aggregation of steps; what matters is the structural issue. To establish the equality of a and c, the one angle, a, is viewed as part of 180°, as part of the angle $a + b$; c is also viewed as part of the —equal—angle $c + b$. With the remainder identical, the angles a and c must be equal. The structural issue in the two equations is:

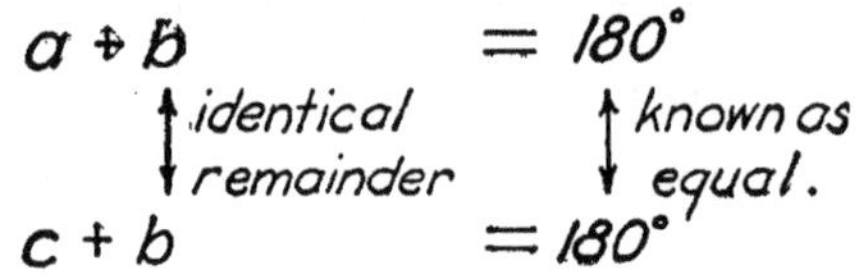

FIG. 53

Thus what matters is how the two equations are structurally related to each other; a sensible procedure looks for these structural requirements. The B-reactions violate them, are blind to them. The A-reactions are determined by them but deal freely with the items; it is immaterial whether the steps in the demonstration are "correctly repeated."

Putting it generally, the structure is:

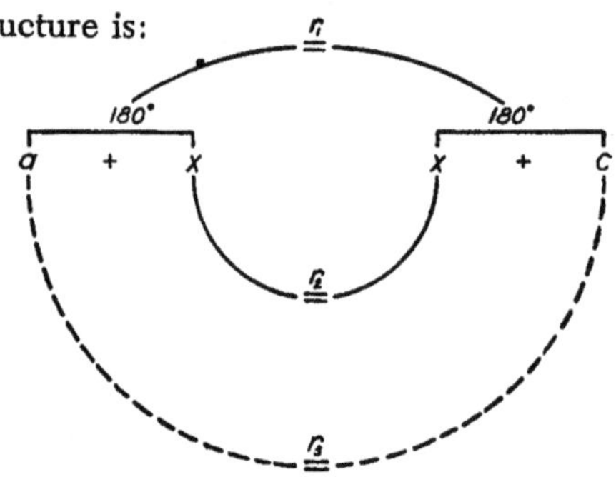

FIG. 54

It is not the items that are decisive but the kind of grouping in connection with the relations: r_1, the equality of the sub-wholes,

r_2, the identity of the remainder,

leading to r_3, the equality of the two angles.

This is not any aggregation of relations or operations: they are interlocked with the task, are sensible parts of a closed whole.

Some theorists recognize that the view of the whole is required but still miss the issue. They formulate certain B-reactions, for example, as follows: "The subject was wrong because he did not take all the items or relations into account." All items? All relations? It is just characteristic of sensible processes not to take all items into account. If this figure is given with the task of proving $a = b$, the fifth line is sensibly neglected. In short, "whole" does not mean "all," but refers to the structure of the items as they must be related in view of the task; it refers to the "good gestalt."

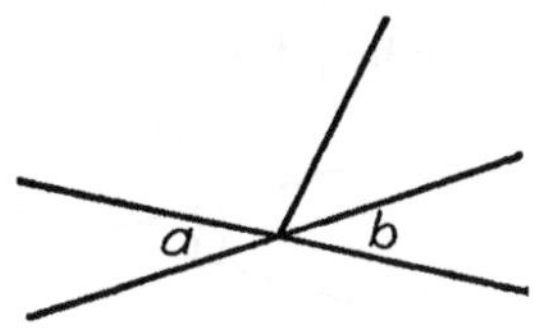

FIG. 55

The reader may get clarification if he applies the structural scheme (p. 85) to the A- and B-reactions. In some of the B-cases—the senseless cases or the desperate cases—one main relation is lacking; in others, there are two main relations, as in

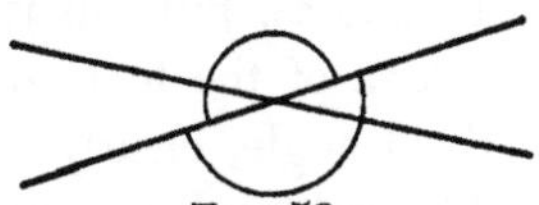

FIG. 56

But the procedures are blind because of the structural place of the items they interrelate. This leads to the conclusion that not the relations alone are decisive, but the relations with regard to their places within the good structure.

In the figure on page 85, relation 1 is a relation not between items but between two groups or sub-wholes, which are seen as symmetrical. Their equality (relation 1) plays the decisive role in the process, whatever the item may be, whether 180°, 90°, etc. Relation 2 is a relation between "homotype" items in the 2 sub-wholes. Relation 1 with relation 2 reasonably implies relation 3 which was sought: $r_1 \; r_2 \supset r_3$. (The logician must not be mistaken about the formula $r_1 \; r_2$ leading to r_3. It is not a case of a transitive relation. The formula is meaningless if the relations are not meant in their structural places.)

The task of finding the proof for the theorem, $a = c$, without help, seems very much more difficult than, e.g., the problem of the area of the parallelogram. Why?

In addition to the point mentioned earlier, that the reason for asking for the proof at all is often ununderstandable, the main reason seems to be that the situation calls for seeing the figure in *two* organizations, *ab/bc*—symmetrical, to be sure, but overlapping, while still tending to the pairing of the two critical angles *a c*.

To view the angle *a* "as playing the same role in *ab* as *c* does in *bc*" demands considerable clearmindedness.[3] There are subjects who help themselves by drawing two figures:

Fig. 57

And in teaching this also seems to help understanding sometimes.

IV

The decisive thing in the A- and B-reactions was the structural relation of the pair of equations. But this is not sufficient. Often in genuine cases the very idea of the first equation, of grouping the given angle with the third angle, clearly comes into existence because it is seen that this can be done symmetrically for both the angles in question. It is not an operation in and for itself, but justified as part of the plan. One feels that the two operations (later equations) will interlock and in this way bring the solution. In such cases there is not simply a succession of two acts, but when the first comes into existence, it does so already *as* one of a pair. It is written down as a separate item but it is not separately conceived.

Thinking does not, as many believe, necessarily proceed merely by passing successively from one item to another, by formulating successive propositions; it does so of course, but in the very act of thinking, in genuine processes, often it does not. Here the procedure goes from viewed whole-qualities to the items viewed as parts of the whole.

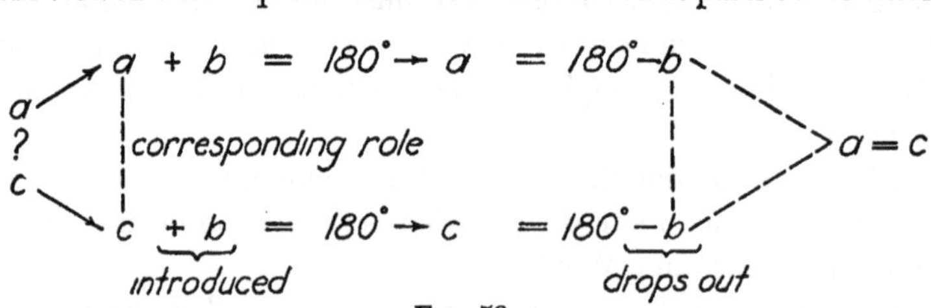

Fig. 58

[3] Often after just sensing the structural requirements, the procedure becomes easier by transition to algebra.

The line of thinking, the direction, is in this case not *one* succession of items; there is a symmetrical two-directedness, each critical angle being dealt with as a part in its whole, formed by the introduction of the third angle, which subsequently can drop out again through the symmetry of the operations.

A comparison: some performances require both hands symmetrically, jointly, to co-operate with each other, to supplement each other's movements. In some cases it would be senseless to proceed with a sheer succession of separate items: you give two playing cards to a child, and ask him whether he can make them stand. The child may take one card and incline it about 30° from the vertical, an action that is sensible only in relation to the idea of the completed structure. To do this with one card, without viewing what will be done with the other, is senseless. There are subjects so educated that they are hampered in their thinking by the habit of proceeding only successively, step by step. But one should not suppose that one always has to do one thing after the other with the idea, "I'll take care of the other things later." Try first to see what you are doing in its context, deal with it as part of the context.

The successive habit—and so the widespread theory that thinking is by nature so[4]—is due to its adequacy in summative situations in which the performance of one operation is merely additively connected with others. It is due further to the fact that we cannot say two sentences at the same time, that we cannot write down two propositions simultaneously, that in reports we have to proceed with one thing after the other. That is one of the reasons why diagrams are often useful.

Further, the sheer successive habit is often caused by the demand for exactness, correctness in every step, which is of course seriously needed but which is not enough. And still further it is due to the fact that correct expression, or logical, formal expression, seemed to be possible only in and-sums of items: to repeat, in connection with the axiomatic assumption that thinking is, must be, verbal by nature, that logic is a matter of language, both of which assumptions are blind generalizations. Viewing the whole seemed unsuited to exact formulation.

[4] Cf. Kant's formulation that human thinking is by necessity *discursive* only.

CHAPTER III

The Famous Story of Young Gauss

I

FIRST a question to the reader:

A staircase is being built along the wall in the hall of a new house. It has 19 steps. The side away from the wall is to be faced with square carved panels of the size of the ends of the steps. The carpenter tells his apprentice to fetch them from the shop. The apprentice asks, "How many panels shall I bring?" "Find out for yourself," rejoins the carpenter. The apprentice starts counting: 1 + 2 = 3; + 3 = 6; + 4 = 10; + 5 = . . .

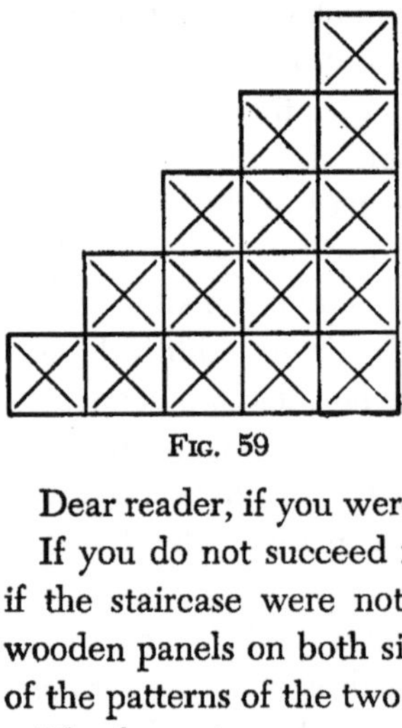

FIG. 59

The carpenter laughs. "Why don't you think? Must you count them out, one by one?"

Dear reader, if you were the apprentice, what would you do?

If you do not succeed in finding a better way, I will ask: "What if the staircase were not along the wall and required the square wooden panels on both sides? Would it help if I suggested thinking of the patterns of the two sides cut out of paper?"

The foregoing are some of the various experimental questions by which I studied features in the range of problems involved in the Gauss task.

Now I shall tell the story of young Gauss, the famous mathematician. It runs about as follows: he was a boy of six, attending grammar school in a little town. The teacher gave a test in arithmetic and said to the class: "Which of you will be first to get the sum of 1 + 2 + 3 + 4 + 5 + 6 + 7 + 8 + 9 + 10?" Very soon, while the others were still busy figuring, young Gauss raised his hand. "Ligget se," he said, which means "Here it is."

"How the devil did you get it so quickly?" exclaimed the surprised teacher. Young Gauss answered—of course we do not know exactly what he did answer, but on the basis of experience in experiments I think it may have been about like this: "Had I done it by adding 1 and 2, then 3 to the sum, then 4 to the new result, and so on, it would have taken very long; and, trying to do it quickly, I would very likely have made mistakes. But you see, 1 and 10 make eleven, 2 and 9 are again—must be—11! And so on! There are 5 such pairs; 5 times 11 makes 55." The boy had discovered the gist of an important theorem.[1] In diagram:

not

but

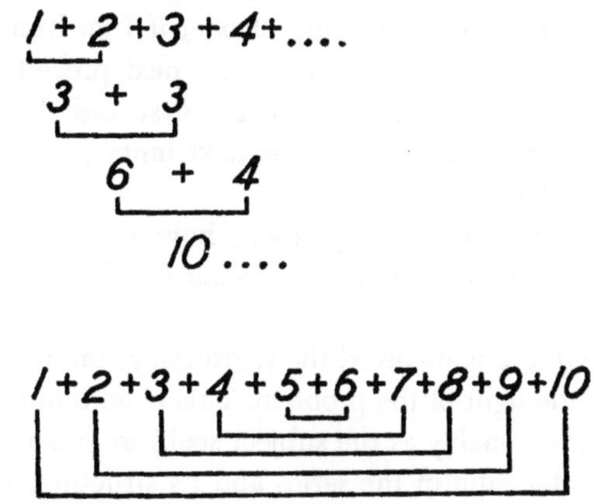

FIG. 60

As the teacher had put the problem to the class, I put it to many subjects, including children of various ages, to see whether a good solution would be found, and what helps, what conditions, might bring it about. In order to study the steps and the features involved, I employed systematic variations some of which I shall describe later. Sometimes I gave very long series. I said directly: "Solve the

[1] $S_n = (n + 1) \frac{n}{2}$

problem without using the cumbersome additions," or I simply waited for reactions.

Here are the best types of genuine processes that I found.

1. At first no way was seen of dealing with the problem. Then: "If a sequence of numbers is to be added, it is certainly correct to add them as they come—but tiresome." Suddenly: "This is not just any sequence; the numbers increase consistently, one by one—this fact may . . . it must have something to do with the sum. But how the two hang together—the form of the sequence and its sum—what the inner relation is between them—is dark, unclear; I feel it somehow but I cannot clarify it."

After a while: "The series has direction in its increase. A sum has no direction. Now: the *increase* from left to right involves a corresponding *decrease* from right to left! This *has* to do with the sum. $\longrightarrow$ more and more; $\longleftarrow$ less and less; in the same amount. If I go from left to right, from the first number to the second, there is an increase of one; if I go from right to left, from the last number at the right to the next preceding, there is a decrease of one. Hence the sum of the first and the last numbers must be the same as the sum of the next inner pair. And this must be true throughout!

"There remains only the question; how many pairs are there? Obviously the number of pairs is one-half of all the numbers; hence of the last number."

Essentially there is involved the regrouping, the reorganization of the series in the light of the problem. This is no blind regrouping; it comes about reasonably as the subject seeks to grasp the inner relation between the sum of the series and its structure. In the process the various items clearly gain a new meaning; they appear functionally determined in a new way. Nine is no longer viewed as 8 plus 1, it has become 10 minus 1, and so on.

If one gets the general formula $S_n = (n + 1)\frac{n}{2}$ in some such way, then one understands its terms in the light of this structure: $(n + 1)$ represents the value of a pair, $\frac{n}{2}$ the number of pairs. But many know only the formula, in a completely blind way. To them all forms $(n + 1)\frac{n}{2}$, or $\frac{n+1}{2}n$, or $\frac{n\,(n+1)}{2}$, or $\frac{n^2+n}{2}$ are simply equivalent.[2] The two *n*'s seem to them to stand for the same item. They do not realize that, in the first formula, n in the expression $n + 1$ is one member of a pair, while n in $\frac{n}{2}$ stands for the number of terms in the series, determining the number of pairs. Of course the four formulas give the same end-result and are in a way equivalent, but not psychologically.[3] In reality they are logically different too if one views them with regard to their form and function, and not solely in terms of external equivalence. Of course it is a matter of logic only if one does not exclude from logic the functional meaning of terms, the genetic question, the question of the approach to the formula—finding or understanding the formula sensibly.

The formula applies equally when the series ends with an uneven number, e.g., 1 2 3 4 5 6 7. Here the grouping described sometimes produces hesitation: what is to be done with the number in the middle which cannot be paired? A further step is required in this kind of procedure. Looking at the single number there comes a sudden discovery: "This must be half a pair, $\frac{n+1}{2}$!" And after some deliberation it is found that this does not change the formula: there are 3 pairs and there is the remainder in the middle, now understood as half a pair.[4]

There are other ways of proceeding productively and sensibly.

[2] Even, e.g., $\frac{(n+\frac{1}{2})^2}{2} - \frac{1}{8}$. Or compare a blind generalization of the formula $\frac{n^2+n}{2}$ into $\frac{n^x+n}{x}$.

[3] The difference shows itself objectively in responses to varied tasks. See p. 96, footnote 7.

[4] $$(n+1)\frac{n-1}{2} + \frac{n+1}{2} = (n+1)\frac{n-1}{2} + (n+1)\frac{1}{2} = (n+1)\frac{n}{2}$$

pair | number of whole pairs | half-pair

The following procedure by an 11-year-old boy is in line with what I have just described. After I had simply put the question to him: "What is 1 + 2 + 3 + 4 + 5 + 6 + 7 + 8 + 9?" he asked, not too well pleased, "Should I count them?" "No," I answered. Suddenly smiling, he said, "There at the end is 9. Eight with the 1 in the beginning is also 9, and so must be the other pairs . . . ," then he stated the result.

2. Another way found by a 12-year-old boy started out differently. The task was: 1 + 2 + 3 + 4 + 5 + 6 + 7.

Told not to do it by counting step by step, he said slowly: "The numbers ascend consistently . . ." And then, suddenly happy, "Oh, I have an idea! I simply take this number in the center and multiply it by the number of terms in the series—which is of course equal to the end number." It was clearly a discovery for him. Asked to show what he meant, he took the middle number, 4, and multiplied it by 7. Given a series ending with 8, he took the middle value between 4 and 5, viz., 4½.

In terms of a general formula, this means

$$c \cdot n \text{ (central value times } n\text{), or } \frac{n+1}{2} \cdot n$$

The formula is structurally different from the first in which $n + 1$ was the sum of each pair, and $\frac{n}{2}$ the number of pairs.

I wanted to become clearer as to what he meant and how he had reached his solution. He was not able to give any clear-cut mathematical formulations, but what he said was: "The numbers ascend consistently. This means that for the sum the central value is significant. The numbers increase to the right of it; and they decrease in the same way to the left of it. So what is added on the right is just what is lacking on the left." In diagram

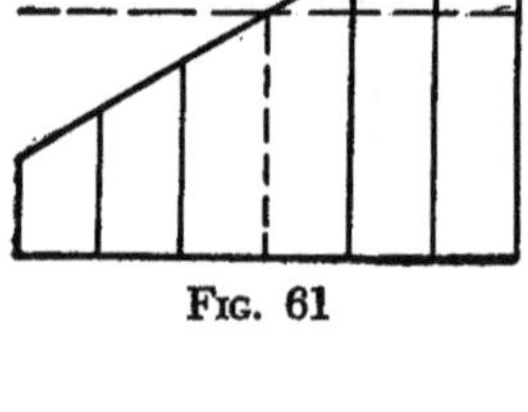

FIG. 61

3. The following was a structurally different procedure: realizing the consistently ascending character of the series, the subject viewed finding the sum as: "Very troublesome because of the jagged edge."

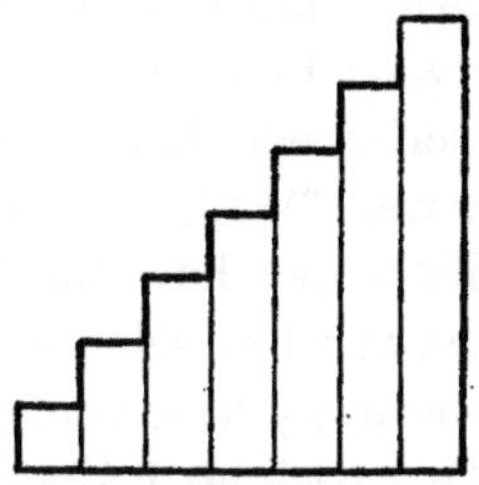

FIG. 62

"But"—and here the subject's face lit up—"I can easily make this trouble disappear. If I combine this staircase with another, inverted one, they will have to fit and give a clear figure without any trouble.

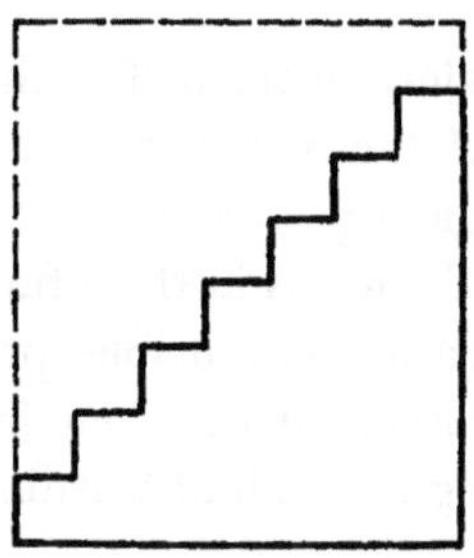

FIG. 63

The sum is clearly base times height, $n \cdot (n + 1)$; half of it."[5]

This provides a sensible foundation for the well-known procedure in which the teacher says, "To get the sum of such a series, write it out, then just write it below in the opposite order and add each vertical pair. They are equal."

$$\begin{array}{l} 1 + 2 + 3 + 4 \ldots\ldots\ldots + 58 + 59 + 60 \\ 60 + 59 + 58 + 57 \ldots\ldots\ldots + 3 + 2 + 1 \\ \hline 61 + 61 + 61 + 61 \ldots\ldots\ldots + 61 + 61 + 61 \end{array}$$

[5] Cf. Chap. I, p. 48 ff. The structural trouble is realized and made to disappear: the two trouble features fit together and vanish in the structurally clear whole.

I found a number of persons who gave this procedure as the solution. They said they had learned it that way in school. Asked why they wrote the series twice, and in this inverted fashion, all were puzzled, did not know what to answer. When I insisted, "What I want is the sum of the series, why find twice the sum first?" for the most part I got the answer, "Well, it leads to the solution in the end." They were unable to say how the idea of doubling might have originated. I confess that I myself was for a long time at a loss to see how one could reasonably have come to the idea of doubling. It had looked to me like a trick, as it does to many, like a chance discovery.[6]

When I showed these results to a mathematician, he said, "Why do you bother about what you call 'functional differences,' 'differences in the meaning of terms'? What matters is the formula, which is identical in all cases."

This attitude is certainly justified if nothing matters but the correctness and validity of the final result. But the moment one tries to get at the psychological process in productive thinking, one *has* to investigate, to view the terms in their functional meanings. These bring about the solution in the sensible, productive processes; they constitute the basic difference between finding the formula in a sensible way and finding it by blind learning or by chance trial and error.

The structural operations in the various procedures which have

[6] Cf. a similar procedure: getting the area of a triangle by doubling it into a parallelogram or, specifically, doubling a right triangle into a rectangle when the area is asked for.

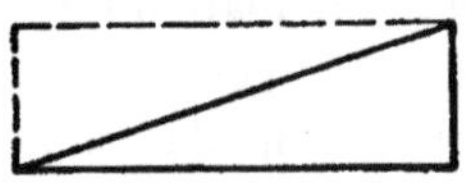

FIG. 64

been described are in some respects different.[7] But there are also characteristic identities: first the problem is seen, realized. It calls for grasping the concrete structure of the series in the light of the problem. The aim of discovering the inner relation between structure and task leads to regrouping, to structural reunderstanding. The steps and operations do not in the least appear to be a fortuitous, arbitrary sequence; rather they come into existence as parts of the whole process in one line of thinking. They are performed in view of the whole situation, of the functional need for them, not by blind accident nor as the thoughtless repetition of an old rule-of-thumb connection.

Although the whole process sometimes takes no more than a minute—as in the case of the two boys mentioned—the idea often comes in a colloidal manner, first in terms of directions, of main lines of groupings, etc. Often it takes some time before the situation becomes really lucid, fully clear. This is especially the case in finding the formula. Having once got the idea, the subjects are able to envisage some of the structural qualities of the equation to be formulated long before they are able to write the formula in concrete terms. That such a period of thought often appears hazy is, I

[7] The organization, the grouping, etc., in the three examples correspond to the following formulas:

$$1.\quad S = \underset{\substack{\text{value of}\\ \text{one pair}}}{(n+1)} \cdot \underset{\text{number of pairs}}{\frac{n}{2}}$$

$$2.\quad S = \underset{\text{center value}}{c\left[= \frac{n+1}{2}\right]} \cdot \underset{\text{number of terms}}{n}$$

$$3.\quad 2S = \underset{\substack{\text{one pair}\\ \text{(or height)}}}{(n+1)} \cdot \underset{\substack{\text{number of pairs}\\ \text{(or base)}}}{n}$$

These differences in reaching the solution also have objective consequences: they even seem easier than the odd series in 1 because of the complication with the half-pair; if procedure 1 has been found or shown in an even series and, as a test, an odd series is presented, difficulties often appear—hesitation, longer reaction time. The opposite seems to be the case in procedure 2: an uneven series seems easier because there is a middle term present, whereas in an even series there is not; if procedure 2 has been found or shown in an uneven series and, as a test, an even series is given, hesitation often occurs. In 3 there is no such difference with regard to even and uneven series; and a long series is often more easily dealt with than in 2, because the center need not be found.

think, due largely to the fact that exact terms for structural qualities, for whole-qualities, are frequently lacking, have not been developed. Of course the matter is really finished only after all the items involved have crystallized. But conceiving of symmetrical compensation is often the essential part in the process. In this phase it often happens that one is clear about rejecting proposed formulas which do not fit the envisaged structural qualities long before one is able to write down the positive formula. This is similar to the state of affairs in which a creative musician has in mind the idea of a melody in its whole-character; he tries to concretize it on the piano, produces something which he definitely discards as not proper, etc., until he does find the tones for what is in his mind.

II

I shall give some examples of the tasks I used in the experimental study of the Gauss problem. As in the problem of the parallelogram, subjects of various ages, especially children, were shown the Gauss method (p. 90) in one example, 1 + 2 + 3 + 4 + 5 + 6, generally without the formula, sometimes with it. Then one or more tasks of the types appearing below were given to see what the subjects would do spontaneously, what help they needed, what kind of help really helped, etc.

The reader may guess the nature of the reactions: what fine productive processes (A-reactions) sometimes occurred (especially with problems *d* and *e*), how subjects sometimes even deepened, broadened the formula—and what blind, stupid B-reactions sometimes occurred.

Let the reader try for himself; let him see what happens to him in solving these problems—in one way or another they are all A-tasks:

What is the sum of:

a. 1 + 2 + 3 + 4 . . . . + 58 + 59
b. 17 + 18 + 19 + 20 + 21 + 22 + 23
c. 1 + 2 + 3 + 4 + 16 + 17 + 18 + 19
bc. 96 + 97 + 98 + 102 + 103 + 104
d. 1 + 5 + 9 + 13 + 17 + 21
cd. 9 + 11 + 13 + 15 + 17 + 19 + 21

What is the product of:

e. 1 · 2 · 4 · 8 · 16 · 32
ec. 5 · 10 · 20 · 40 · 80 · 160
f. ⅛ · ¼ · ½ · 1 · 2 · 4 · 8

I have said that these are all in some way A-tasks. I hope you see that.

In a) the original series is prolonged. If the formula has been taught, it is simply a special case of the formula.

b) does not start with 1. How did you deal with it? Did you see any direct way? Of course I made it easier for you by choosing a round number, an outstanding number. Did you think of an inclusive formula?

Series c) has a gap. Did it disturb you?

In series d) the difference between the terms is changed. What did you do?

Series e) and f) ask for the product. Were you surprised? Did you find a way? Could you write the formula?

I did not of course teach formulas to young children, nor did I ask for them; I often chose simpler numbers than in series b) and bc), or simpler cases than series e), f), but not necessarily shorter series, often much longer ones. One must be careful about the relation of successive tasks to each other. The nicest thing is to jump from the original to one of the last problems, to d) or e).

Often one has interesting experiences with such tasks: sometimes one gets surprisingly fine reactions, which are also evident in the remarks of the subject; sometimes one encounters utter helplessness, surprisingly stupid or blind responses even in intelligent subjects, especially if they have been blinded by habitual attitudes or by drill (cf. Chap. I, p. 18). The nature of both the sensible and the senseless reactions illuminates the psychological issues.

For problems of type e), f), requiring the transition from addition to multiplication, I may cite the following incident: I had shown the Gauss method, in the example 1 + 2 + 3 + 4 + 5 + 6 + 7 + 8, to a boy of eleven years. Then I gave him the series 1 · 2 · 3 · 10 · 15 · 30. "No," he said, "here it is impossible to use that nice method . . . ," but after a while he added spontaneously: "Oh, if I were to multiply these numbers it would work . . . !" and he showed the grouping 30 · 30 · 30, making the discovery of the application to products by himself.

In the addition form this last series was a B-case; the multiplication form was an A-case. It became possible in experiments to use systematically pairs of A- and B-forms of such series as the following:

$5 + 10 + 20 + 40 + 80 + 160$	(B-case)
$5 \cdot 10 \cdot 20 \cdot 40 \cdot 80 \cdot 160$	(A-case)
$1 + 2 + 4 + 8 + 16 + 32$	(B-case)
$1 \cdot 2 \cdot 4 \cdot 8 \cdot 16 \cdot 32$	(A-case)

On the other hand, in some series the problem in addition form was an A-case:

$5 + 10 + 15 + 20 + 25 + 30$	(A-case)
$5 \cdot 10 \cdot 15 \cdot 20 \cdot 25 \cdot 30$	(B-case)

Or:

$1 + 2 + 3 + 4 + 5 + 6$	Original case
$1 \cdot 2 \cdot 3 \cdot 4 \cdot 5 \cdot 6$	(B-case)
$1 \cdot 2 \cdot 3 \cdot 4 \cdot 6 \cdot 12$	(A-case)
$1 + 2 + 3 + 4 + 6 + 12$	(B-case)

Which cases one rejects, to which ones one applies the method, what difficulties one has, etc.—these things are characteristic for understanding.

There are similar examples of B-tasks more liable to induce blind reactions. If, for example, we give, instead of series

a) $1+ 2 + 3 + 7 + 8 + 9$,
the series *b*) $1 + 2 + 3 + 4 + 7 + 8 + 9$
or *c*) $1 + 2 + 3 + 4 + 6 + 7$,

subjects sometimes show blindness to the requirement of the symmetrical position of the gap. Some subjects, on the other hand, apply the method correctly and unhesitatingly (A-reactions) in tasks of type *a*, while in forms of type *b* and *c* they hesitate, even though in piecemeal terms these types are certainly *more* similar than type *a* to the original series $1 + 2 + 3 + 4 + 5 + 6$. They discriminate strongly between the types, looking for the required symmetry and, for the most part, they develop an appropriate, more complex procedure, e.g., restoring symmetry in *b* by eliminating the 4, adding the missing 5 in *c*, or changing the 4 to 5, etc.

Other examples of A-B pairs for distinguishing the issues in type d) are the following:

	$1 + 2 + 3 + 4 + 5 + 6$		
A	$3 + 5 + 7 + 9 + 11 + 13$	A	$1 + 3 + 5 + 7 + 9 + 11$
B	$1 + 2 + 3 + 4 + 11 + 13$	B	$1 + 2 + 3 + 7 + 9 + 11$

Although it does take a considerable degree of blindness, the B-case, especially when it is longer, provides occasion for blind application of the Gauss method if the subject acts really thoughtlessly. On the other hand, B is often intelligently rejected, or dealt with by the cumbersome method, while A is dealt with sensibly.

One can in such ways realize, study and test the structural features involved in the Gauss task, those which are "essential," the features of the structural inner relatedness between operations and form, in contrast to other factors which are structurally peripheral. In the various types of tasks the issues were:

in b), independence of the structural factors from the place of beginning;
in c), the necessary symmetry of the series, tested by means of the gap and its position;
in d), independence of the structural features from the amount of the constant difference;
in e), independence of the structural inner relatedness from the nature of the particular operations, shown by the transfer to the structurally similar cases of multiplication.

It is most interesting to study which forms of tasks are better for discovering the method with or without help; and it is most illuminating for the theoretical problems to find, for example, that shorter series are by no means always more favorable for discovery, nor even that $1 + 2 + 3 + 4 + 5 + 6$ is necessarily more favorable than, e.g., $1 + 3 + 5 + 7 + 9 + 11$.

One primitive factor should not be forgotten: a disordered series, with its numbers mixed up, offers particular difficulties, whether in application or in discovery. The proper order makes for one comprehensive view, shows the needed consistency of the series. On the other hand, there are changes of order which do not seem so unfavorable. What seems important is not the amount of piecemeal deviation from the original; it is rather the kind of order which may help or hinder a clear view of the whole. In

$$1 + 10 + 2 + 9 + 3 + 8 + 4 + 7 + 5 + 6$$

it sometimes happens that a subject stops, exclaims, "It goes right on consistently—these go up, these go down," showing

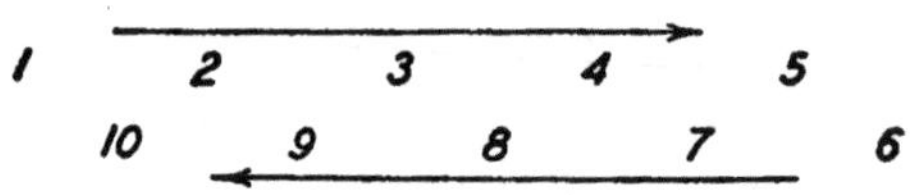

FIG. 65

or making the pairs,

FIG. 66

The latter procedure borders on the well-known "short cut" procedures used by many bookkeepers in long additions. They look for pairs or triplets, making easy round numbers rather than follow the order of succession. These procedures, of course, lack the fine relation to the "principle" of the series.

III

Confronted with the problem of the sum of a series (cf. §1), and given no help, many fail to find the Gauss solution. Why? What makes the task so difficult for many? What does it mean when one says, "It took the genius of young Gauss to do it"? Or why was it that the young boys in the examples mentioned, did it, and did it consistently and easily? What is at the bottom of these achievements psychologically?

The Gauss tasks involve structural difficulties. To overcome them, really to see one's way in spite of them, requires something. On the basis of my experiences, I would say that the essential features in genuine solving are:

not to be bound, blinded by habits;
not merely to repeat slavishly what one has been taught;
not to proceed in a mechanized state of mind,
in a piecemeal attitude,
with peacemeal attention,
by piecemeal operations;

but to look at the situation freely, open-mindedly, viewing the whole,
trying to discover, to realize how the problem and the situation are related,
trying to penetrate, to realize and to trace out the inner relation between form and task; in the finest cases getting at the roots of the situation, illuminating and making transparent essential structural features of regular series, in spite of the difficulties.

The Gauss situation *is* structurally complex; the main difficulty seems as follows: envisaging the inner relation between form and task (sum) is difficult (1) because the compensating differences are hidden

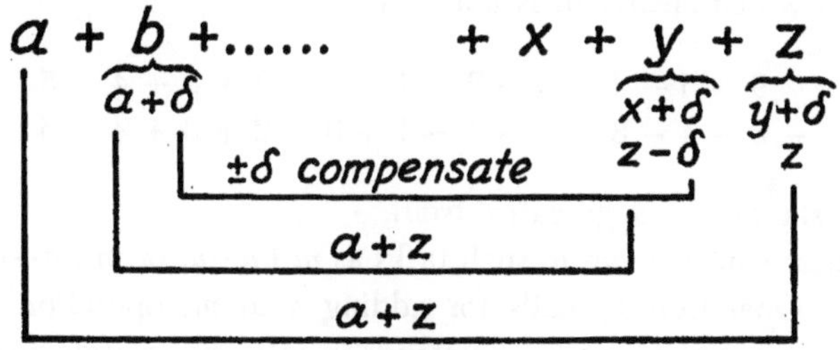

Fig. 67

and (2) because of the psychologically strong form of the ——→ progressive series which has to be broken down into the required symmetrical halves ——→ and ←——.

What if we simplify the structure of the situation, not just by giving fewer numbers, but by employing tasks in which the basic structural features are not so hidden?

Some forms of tasks similar to the previous examples clearly made matters easier, for example:

$$99.8 + 99.9 + 100 + 100.1 + 100.2 = ?$$

$$273\tfrac{3}{5} + 273\tfrac{4}{5} + 274 + 274\tfrac{1}{5} + 274\tfrac{2}{5} = ?$$

or

$$\frac{271 + 272 + 273 + 274 + 275}{5} = ?$$

But let us proceed in a radical way. Let us use tasks in which the compensating differences are not hidden structurally. The achievement becomes natural if, for example, we ask what is the sum of $-3 - 2 - 1 + 1 + 2 + 3$.[8]

Of course there are some who still stick to the attitude of drill, who still proceed blindly, piecemeal. But most subjects, looking at the whole line, laugh when they are given such a series, or are puzzled by the important-looking appearance of so obvious a problem; this happens often with virtually all subjects. In such cases it sometimes happens that one gets the response *without* even putting the question, *without* asking for the sum. If the series is long, it is often plain that the thing is done not by actually forming all the individual pairs, but by a view of the characteristically structured whole, based on the consistent progression of the series. If an item is added which clearly does not fit as, e.g.,

$$9 - 5 - 4 - 3 - 2 - 1 + 1 + 2 + 3 + 4 + 5$$

or

$$-5 - 4 - 3 - 2 - 1 + 1 + 9 + 2 + 3 + 4 + 5$$

it often stands out, segregates itself.

Our case comes down to such tasks as $m+a-a$; or $m+a-a+b-b+c-c$. Operation 1) calls for adding a to m; operation 2) for subtracting a, but operation 2) stands in intrinsic relation to operation 1), being its opposite. Operation 2) appears in this context as demanding the undoing of what would be done by operation 1); and vice versa. That is their structural meaning. Both are viewed and function not as an and-sum of two operations but in their inner relation which makes it unnecessary, indeed senseless, to perform either.

Natural, sensible grasping is at work in the realization of this

[8] Cf. also example *f*, p. 98. Did the reader see it more readily than *e*, *ec*, or even *d* and *cd*?

relation, in the reluctance to perform things which undo each other. The trained psychologist may think of behavior tendencies even in rats. It seems very difficult, often even impossible, to teach a rat maze runs parts of which undo each other.

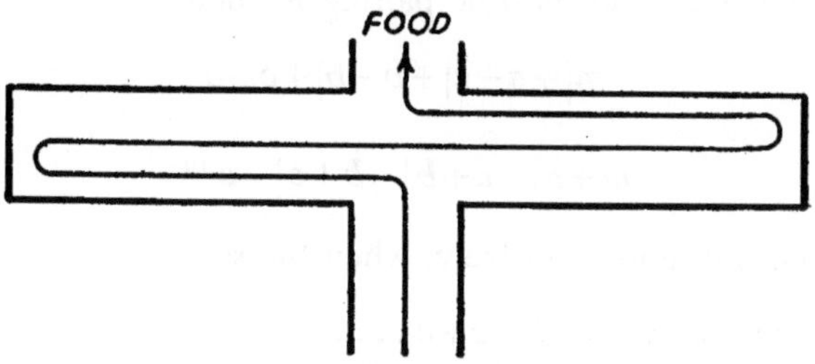

FIG. 68

We must not forget, on the other hand, that there are situations, forms of performance, in which a kind of doing and undoing becomes sensible —in rhythmical play, in rhythmical dance as in $-1+1$, $-1+1$, etc., or in $-1+1$, $-2+2$, $-1+1$, $-2+2$, etc. The symmetry of the opposite movements plays an important positive role.

In 1931 in the Frankfort Institute I gave Miss Siemssen the task of studying psychological differences between sensible and senseless work. As compared with putting books in order, putting them sensibly on library shelves, we used externally similar Sisyphus tasks: rows of books were to be put on the shelves, then taken out again and replaced in their former places, put on the shelves again, and so on. . . . In both cases the activity was observed for about half an hour. The subjects performed the senseless task politely enough, although reluctantly, and obviously with some difficulty in sticking to it; as time went on, resistance increased and revolutionary tendencies came to the fore. But sometimes, when the performance continued, a striking result emerged: with some subjects the character of the task changed into something nicer—it became a kind of rhythmical dance, the books were taken out and returned in measured dance movements; continuing then was no longer so hard, the task had changed into jolly play. Even this did not last long, however.

To return to the issue: the role of sensible order, the features of sensible grouping in the situation, become technically clear if we give children tasks of the following types, and compare their attitudes and responses:

1. $m+a-a+b-b+c-c$
2. $m+a+b-c-a+c-b$

3. $m+a+b+c-a-b-c$

or 4. $m+a+b+c-c-b-a$ etc., with or without the $m-$item.[9]

In the first we get easy, quick responses for the most part: "Of course it is m," with occasional remarks such as, "Should I always undo what I do?" The sensible pairing is done

$$m|+a-a|+b-b|+c-c$$

never

$$m+a|-a+b|-b+c|-c \text{ [10]}$$

Similarly, but more decisively, when the series is

$m-a+a-b+b-c+c \ldots.$

one gets $m|-a+a|-b+b|-c+c \ldots.$

not $m-a|+a-b|+b-c|+c \ldots.$

Most subjects do not bother to make the addition of $m+a$ or the subtraction $m-a$. Or, if they do, they are soon annoyed with themselves, exclaiming, "How stupid, how blind!"

In the second task we find more piecemeal, blind procedures. Often there is hesitation, uneasiness; remarks such as, "This needs to be put in order," "It is out of order," and the children rewrite the series in sensible pairs.

The third type of task again seems easier than the second, giving a quick view of the corresponding halves; the series is still more

[9] Also concrete cases as, for example:

96 + 77 − 77 + 134 − 134

or 96 + 77 − 134 − 77 + 134

or 48 + 79 − 124 − 79 + 124

or 48 + 79 − 79 + 124 − 124. The last in blind procedure:

48 + 79 = 127

127 − 79 = 48

48 + 124 etc.

[10] It is illuminating for the theoretical views of the transfer problem, to consider A-B cases in elementary forms:

1) Original shown,	taught: $a+b-a$	e.g.,	35 + 14 − 35
2) A−form	$c+d-c$		87 + 69 − 87
3) B−form	$a+b-c$		35 + 14 − 87
4) A^1−form	$a+b-b$		35 + 14 − 14

In 1) the procedure of grouping the first term with the last is "shown, taught." In 2) all the numbers are changed as compared with the original example. In 3) much less is changed; viewed piecemeal, in terms of sum and stimuli, it is much more similar to the taught original. But if any understanding has taken place, the transfer is made by children to tasks 2) and 4), but not to task 3).

easily grasped if the given digits are not arbitrary, but a surveyable principle is used as in $m-1-2-3+3+2+1$, and in other similar examples.

A simple experimental technique for the study of these sensible tendencies of grouping is what we may call the "square set." The task is to add 4 numbers, of which the addition of two gives an outstanding number or mutual compensation.

$$1)\quad \begin{array}{r} +a-a \\ -b+b \end{array} \qquad\qquad 2)\quad \begin{array}{r} +a-b \\ -a+b \end{array}$$

$$\text{e.g., } 1)\quad \begin{array}{r} +56-56 \\ -27+27 \end{array} \qquad\qquad 2)\quad \begin{array}{r} +56-27 \\ -56+27 \end{array}$$

Set 1) is usually sensibly conceived and dealt with in horizontal pairs, set 2) in vertical pairs. Similarly, with tasks in which two or more numbers make an outstanding number rather than compensate each other:

$$1)\quad \begin{array}{r} +98+2 \\ +75+25 \end{array} \qquad\qquad 2)\quad \begin{array}{r} +98+75 \\ +2+25 \end{array}$$

If we designate the 4 terms in all such sets by $\begin{smallmatrix} a & b \\ c & d \end{smallmatrix}$ the preferred grouping in sets of type 1) is *a b/c d*; in sets of type 2) *a c /b d.* The psychologist sees that these are cases belonging to the results of investigations about the role of organization in perception, which uncovered the so-called "gestalt tendencies" in grouping.[11]

[11] Cf. M. Wertheimer, "Untersuchungen zur Lehre von der Gestalt," *Psychologische Forschung*, 1923, Vol. 4, pp. 322-323; see also W. D. Ellis, *op. cit.*, p. 82. For example, Fig. 69 is seen as *ad/bc*, not as *ab/cd*. Again, Fig 70 is

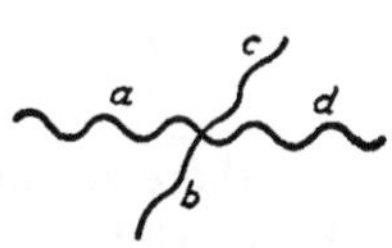

FIG. 69

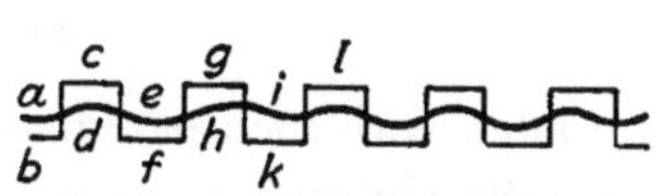

FIG. 70

seen as *bcfgkl . . ./adehi* . . . , not as *acegi . . ./bdfhk* . . . ; it is nearly impossible to realize the latter view of the whole figure in perception.

In those experimental investigations, dealing mostly with point sets or simple figures, it was an illuminating moment when a strong tendency was found to perceive consistent whole-qualities, "reasonable groupings" with features belonging to the inner structural nature of the situation—the so-called factor of the "good gestalt."

Those investigations showed that the tendency of perception to be "reasonable" is correlated with sensible, lawful mathematical features of the situations—although with restrictions, because of the importance of whole-qualities rather than of "class-laws" in perception (cf. p. 115 ff.).

The issues with which we are concerned here, are by no means peculiar to arithmetic or to arithmetical training. A figural example not unlike the arithmetical square sets is the following optical constellation; especially in solid figures, e.g., black figures on white.

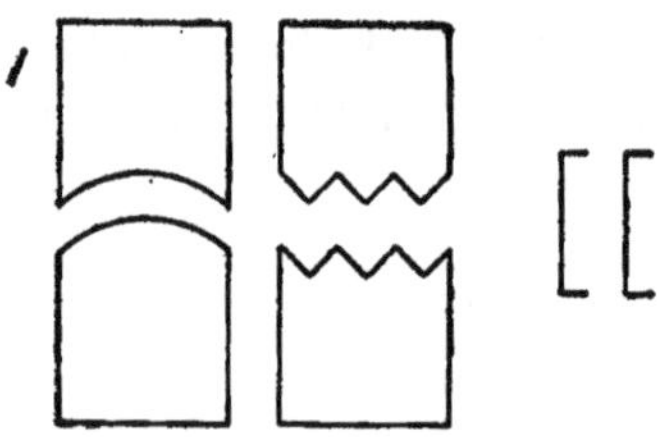

FIG. 71

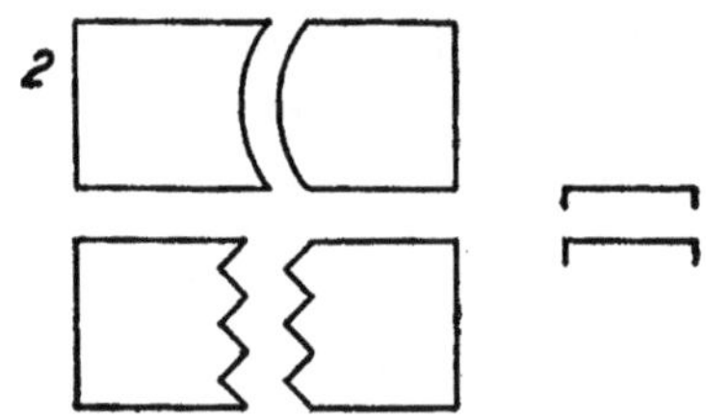

FIG. 72

Set 1) is usually seen vertically paired, set 2) horizontally.[12]
Or take these situations:

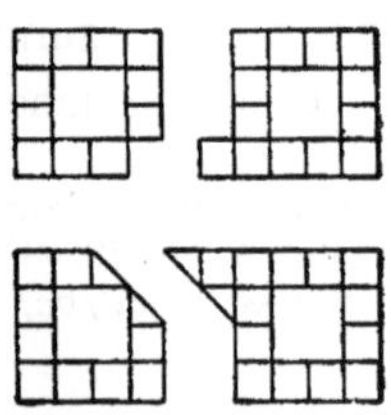

FIG. 73

If confronted with such patterns—say in blocks—there seems to be even in little children a strong tendency in the sensible direction.

[12] Cf. experiments on movement with the special method of square sets. E.g., P. v. Schiller, "Stroboskopische Alternativversuche," *Psychologische Forschung*, 1933, Vol. 17, pp. 179-214.

They often get at it spontaneously, "improving," "correcting" the situation. No language is needed—they just put the objects together reasonably, fit them together. Often it is not even necessary to assign a task for sensible response to appear: it grows out of the inner dynamics of the situation. Again we see the role of "disturbance," "gap," "fitting," "being just needed," "being required," as parts in a consistent whole. These features seem also to be of basic importance for the reasonable teaching of arithmetic.

A simple illustration for our problem is this figure, which reveals[13]

FIG. 74

a tendency to take away the square, or the remainder, where it is "too much," and to put it in the place where it is lacking.

Similar considerations seem to be of basic importance in teaching geometry. For example, even to grasp sensibly, what an *angle* is, it seems important to view it as part in the outstanding whole of 360°. If angles of 182° and 180°, 355°, 360°, 363°, are dealt with

[13] In many years of study with children, Dr. Catherine Stern has developed tools and methods for the teaching of arithmetic in which genuine discovery in tasks of a structural nature plays an essential role. The results in learning—and in happiness—seem extraordinarily good as compared with the usual teaching by drill which focuses on forming associative bonds, etc. These methods and studies will be published soon.

indifferently as just *any* angles, as of the same rank, one is blind to their structural place, to their functional meaning. Here I will mention experiments with children who were asked to turn the large hand of the clock by several angles in succession.[14] The tasks were similar to the Gauss problem. For example: what is the end position of the hand if it is turned clockwise, first through 7°, then through 90°, then 180°, then 90° again? Or first through 8°, then 7°, then 83°, 6°, 84°, 5°, 85°, 4°, 86°? In experiments with children who did not know about angles, I said, "It is 12 o'clock; suppose I turn the hand several times, clockwise. Where will the hand be in the end if I turn it first 7 minutes, then 25, 5, 24, 6?"

Here is an experience which occurred in dealing with the following type of task with grownups. I asked for the sum of the vectors, the forces working on a body, in instances like the following: "There is one vector (a) working vertically upward (0°), size K; another (b) at an angle of 90° to it, size L; a third (c) at 180°, size K; a fourth (d) at 270°, size L. What is the result of these forces on the body?"

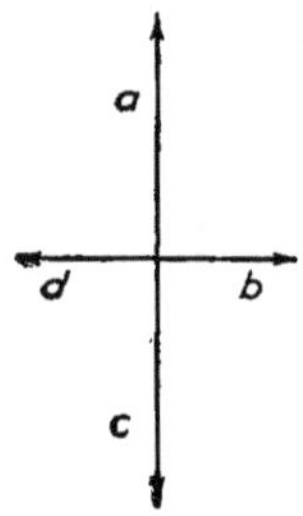

Fig. 75

Now, especially if one makes the diagram and looks at it, the result is clear—zero; the opposites compensate each other, the opposite equals are paired.

[14] See M. Wertheimer, "Über das Denken der Naturvölker, Zahlen und Zahlgebilde," *Zeitschrift für Psychologie*, 1912, Vol. 60, pp. 321-378; reprinted in Wertheimer's *Drei Abhandlungen zur Gestalttheorie* (Erlangen, 1925); see also W. D. Ellis, *op. cit.*, selection 22.

But it happened that one person, seeing the complete picture, insisted upon doing it in what he called "the exact way." Making the constructions, he said, "Vectors a and b in the parallelogram of forces give the resultant r_1.

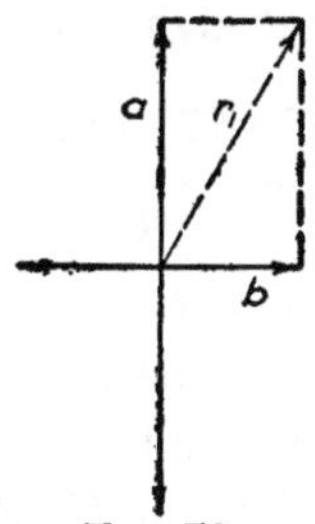

FIG. 76

Combination of the first resultant and vector c in a parallelogram of forces gives the second resultant. This with d gives resultant 3, which is zero; and r_3 combined with a, gives $+a$ as the result." He was clearly puzzled, said hesitatingly, "But this is nonsense! And yet, working it out, it comes out that way . . . where is the mistake?" He did more than 14 minutes of strenuous thinking without becoming clearer, and dropped the matter. Coming back to the problem after a time, he suddenly said rather sadly: "I've got it. I had already used the first vector," and in an apologetic tone he added, "I was dumb. It was clear to me that I had to go all around. With the resultant 3) I felt that I had gone only ¾ of the way, only as far as 270°. . . . I thought I had to complete it. Didn't think that I had used vector a already. How stupid I was. Of course, a and c make zero, so do b and d. So the resultant is zero."

FIG. 77

Certainly his procedure was correct up to the last step. Construction of every resultant is often needed, it is the general method. But we should not forget that often, in productive situations, the sensible view of the whole plays a decisive role: the realization of symmetry and balance in the whole, and of deviations from it, in sensible grouping. He was obviously misled by the closing tendency which obtruded.

To be sure, this was an extreme case. If the diagram is shown or drawn, almost all responses are sensible—always assuming that the meaning of "vectors" is clear.

IV

I have mentioned before that giving the task in the form of

$$\frac{271+272+273+274+275}{5}=?$$

seemed helpful. Some see their way at once. "This is 273, of course," they say and do not even start the cumbersome additions. Some do not see it, ask whether they really should do all the additions. If the task is given as a test after the Gauss's method has been taught, the subject may even then start blindly with "271+275=546."

In this form the point is that the denominator requires the numerator to be divided into five equal parts and so helps toward viewing the expression above as being these five parts. When experiments showed that the concrete difficulties of many subjects with such tasks were similar to those of the Gauss problem, structural simplification seemed to be the next step.

When I asked children,

$$\frac{274+274+274+274+274}{5}=? \text{ or } \frac{272+272+272}{3}=? \text{ or } \frac{273+273}{2}=?$$

I got clear-cut results with some bright subjects. Most of them laughed, enjoyed the joke, while others were puzzled that such an easy problem should be given, or were bored; but they had no difficulty with the answer. They realized spontaneously and easily that what the denominator demands is already done in the numerator. Division by five was understood in its structural meaning, as demanding structurization of the amount of the numerator into five equal parts, which is already done. Or, the numerator, viewed as multiplication, showed the compensation of multiplication and division.

To add or really to multiply in order to divide later corresponds here to the doing and undoing we had before; it means working elaborately on something that is already done, trying to reach a solution that is given. Of course, something is to be done, viz., to realize that the solution is already there, to see that one of the numbers is not merely something to be added to the others but *is* the solution. This is an achievement: in the context of the task, intelligent change of the functional meaning of an item into the solution. But that is quite easy; there is no strong "embedding" to be overcome.[15] Although there was sometimes a little hesitation because the subjects had not expected such an easy task, a smile soon appeared with such remarks as, "It is obvious. At first it appeared to be a difficult task but it isn't," and the solution was given.

Thinking of certain school attitudes I had so often experienced, I continued to ask such questions. Quite a surprise was in store for me—I had not imagined how extreme the situation often was. A number of children who were especially good in arithmetic in their school were entirely blind, started at once with tedious figuring or begged to be excused from the cumbersome task—they did not look at the situation as a whole at all. To be sure, when I helped them to see, they were clearly ashamed, exclaiming, "How dumb I was! How stupid!"

These experiences reminded me of a number of more serious experiences in schools, which had worried me. I now looked more thoroughly into customary methods, the ways of teaching arithmetic, the textbooks, the specific psychology books on which their methods were based. One reason for the difficulty became clearer and clearer: the emphasis on mechanical drill, on "instantaneous response," on developing blind, piecemeal habits. Repetition is useful, but continuous use of mechanical repetition also has harmful effects. It is dangerous because it easily induces habits of sheer mechanized action, blindness, tendencies to perform slavishly instead of thinking, instead of facing a problem freely.

[15] One has characteristic experiences in experiments with tasks in which the solution is factually contained in the text of the problem, but functionally strongly embedded, i.e., given in a definitely different functional place and role within the context of the question. Subjects are often blind even to the exact word-for-word formulation of the solution in the text, and characteristic processes occur when, after a time, they discover it. This again furnishes experimental proof of the importance of realizing the structural place, role, function of an item in its whole. (Cf. N. R. F. Maier's experiments on embedding technical tasks, "Reasoning in humans. I. On Direction." *Journal of Comparative Psychology*, 1930, Vol. 10, pp. 115-143.)

An investigation of the blinding effects of mechanical repetition in sequences of assigned tasks was begun in the Berlin Institute in 1924. Duncker and Zener obtained striking results.[16] In recent years, a student of mine, A. Luchins,[17] has made a comprehensive investigation in schools of this effect, and has further developed experimental methods for dealing with it. It is surprising how easy it is, through some kind of mechanization, to blind even bright, highly educated subjects with the beloved method of repetition. Luchins also employed methods of recovery from blindness thus induced, which usually make it easy to regain sensible reactions but which showed no considerable positive effect with many of the children in certain schools. Of course there are a number of possible explanations to be considered both for the blinding effect and for the failure to recover; Luchins and Asch[18] made experiments concerning such theoretical issues. This much seems clear: important factors are the habits developed by drill, the attitude toward problems, certain school atmospheres as to learning, doing, and thinking.[19]

I will now report three reactions to these findings.

I once told a famous psychologist of these results. I said that they might have been due to bad schooling, a consequence of the emphasis on thoughtless association and drill, which impairs attitudes toward thinking. "Oh, no," he said, "not at all. The negative result is not at all surprising if you ask such gestalt questions; the children are not taught to solve such problems. They learn arithmetic in school. If you were to teach them such gestalt tasks, they would learn them. It is only a matter of what they are taught."

These remarks illuminate the theoretical situation. This psychologist is himself a fine thinker. His remarks become understandable if one realizes that, theoretically, thinking is for him, as for many others, fundamentally nothing but the working of mechanical associative bonds, habits, acquired by repetition. What else *could* it be?!

A mathematician whom I told of these experiences remarked: "You are wrong. To see the way to such a short cut is not important;

[16] See N. R. F. Maier, *op. cit.*

[17] A. Luchins, "Mechanization in Problem Solving: the Effect of Einstellung," *Psychological Monographs*, 1942, Vol. 54, no. 6.

[18] S. E. Asch, "Some effects of speed on the development of a mechanical attitude in problem solving," paper read at the 1940 meetings of the Eastern Psychological Association.

[19] As to the consequences of structurally blind teaching, compare Chap. I, section II; compare also the results of Dr. Katona in *Organizing and Memorizing* (Columbia University Press, 1939).

the method of counting out, thoroughly, is a correct method, the general method. You can use the short cut only in specific cases."

Now this is a serious question. In my answer I first referred to several things of which I have spoken in previous chapters. Then I asked him whether he regarded the Gauss discovery too as merely an unimportant short cut. But, thirdly, I said: "I regard it as just the opposite, not a matter of a short cut merely, of a special case. The basic attitude toward a problem, toward an operation, is at stake. For many school children division *means* the drilled technique as, e.g., in $\frac{816}{3}$: '3 into 8 is 2; carry the 2; 21 by 3 is 7; 6 by 3 is 2 . . . 272.' This *is* to them division. But though mechanical mastery is of value, especially in freeing the mind for more important tasks in a problem situation, it should not blind the subject. It makes a difference whether the technique of division is viewed, known, applied as a mere technique or whether a subject is blind to the very meaning of division: the subdivision of an amount into equal parts in all the concreteness of the given structure. And so with multiplication.

"If, in such cases, one is unable to realize the structural meaning of division, one is blinded to the fundamentals. I mean it seriously: teaching arithmetic should not mainly stress drill but should let the child discover the structural features and requirements of given situations, and let him learn to deal with them sensibly. This of course requires a very different procedure from the drill of most schools." I then told him of some developments, especially of Dr. Catherine Stern's[20] structural methods, which of course he enjoyed.

The reaction of another well-known psychologist was different. After I had told him briefly of my experiments in schools, he said: "Of course, I see. It reminds me of an experience I had some months ago which may be typical. My son, who is a bright boy, came and said to me, 'You see, daddy: I am very good in arithmetic at school. I can do addition, subtraction, multiplication, division, anything you like, very quickly and without mistakes. The trouble is, often I don't know *which* of them to use. . . .' "

It is not the teachers who are at fault. Many are somehow deeply dissatisfied with the emphasis on mechanical associations, on blind

[20] Cf. Chap. III, p. 108, footnote 13.

drill. Many rely on them because they seem to them to be in line with scientific psychology—by which they mean the psychology of learning rote syllable series, and of conditioning. Many rely on them because they see no other, more sensible, concrete, scientific way of teaching. To develop better methods is really the task of a more adequate psychology of thinking and learning.

V

Probably the reader now has a clear idea of the psychological structure of the Gauss problem. One interesting point has not received enough consideration in the variations reported. It is just the point that made the Gauss discovery so beautiful: its inner connection with the principle of the series. In the course of experiments I showed a series of numbers without giving a task. Here is one:

$$-63,\ -26,\ -7,\ 0,\ +1,\ +2,\ +9,\ +28,\ +65$$

Perhaps the reader, looking at the series, has already done something with it. He may, as some do, have noted the similarity of certain numbers (-63, $+65$; -26, $+28$; -7, $+9$), stated that the sum of each pair is two; 3×2 is 6; and the sum of $0+1+2$ is 3; so that the sum of the series is 9. This procedure is somehow Gaussian and yet not quite.

There is another type of reaction. I report a typical protocol: "To the right the series increases progressively; similarly it decreases on the left. These numbers correspond somehow, the -63 and 65, -26 and 28, -7 and 9. What about the middle region?

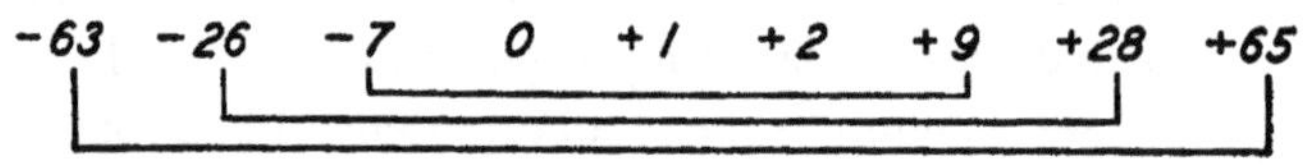

FIG. 78

. . . Oh, the series is out of place!! The real center is where the $+1$ is! This 1 should be zero . . . And if we subtract 1 throughout we get $x_n = n^3$."[21]

Now, such a procedure was also followed in a case in which the

[21]

-64	-27	-8	-1	0	$+1$	$+8$	$+27$	$+64$
(-4^3)	(-3^3)	(-2^3)	$(-1)^3$		1^3	2^3	3^3	4^3

sum was asked for from the start. Interested in grasping the series, however, the subject at first disregarded or temporarily forgot this task. After the subject in this way had reached "It is $x_n = n^3$" he was reminded that the task was to find the sum. "The sum?" he said. "The sum of such a series is naturally zero. . . . Oh, forgive me, here is this silly shift. The whole series is shifted by +1. Every number gets an additional +1. So +1 times the number of terms . . . how many are there? It makes 9." He said it not too pleasantly.

Here someone remarked: "What queer behavior! You were asked for the sum; why did you bother with these things at all?" And he showed the short procedure mentioned above, adding: "No principle was asked for. Why not do the task the direct way?"

To which the first subject, somewhat absent-mindedly, seemingly preoccupied, somewhat annoyed, answered: "Oh, yes, yes,—you are right—but please don't disturb me. Don't you see what opens up here? . . ." He was submerged in thought. There started for him a long process, a chain of discoveries.

To focus on a set question, to solve it in the shortest way, is not always the most intelligent attitude. There is such a thing as trying to get at the root of a situation. Days afterward this same subject remarked: "That silly shift—I just had to see through it." It is a fine thing to uncover the "real" structure,[22] to see through a misleading appearance, to get at the core of a situation, to realize what one is facing.

After a while the subject said: "Here it was $x_n = n^3$ . . . the sum is zero whether the series is symmetrically continued or cut off at any particular point. This would not be the case in $x_n = n^2$. There is equality of the two halves but they do not compensate: $(-2)^2 = +4$, the same as $(+2)^2$. In general, when the exponent is an uneven number, the sum must be zero." After a while he proceeded: "The same will be true of continuous curves, e.g., of the sine curve,

[22] In order to make reasonably sure whether such a structural view, here $x_n = n^3$ (shifted), is the true view in an actual case, one looks ahead to see whether or not further values, left and right, would be in accordance with the envisaged principle. One also examines what happens to the values in variations of the series. But all this is not the issue in this case in which the subject was intent upon certain whole-qualities of consistent series, as his further procedure showed.

properly cut; of the area or of the sum of the vertical lines between the sine curve and the axis:

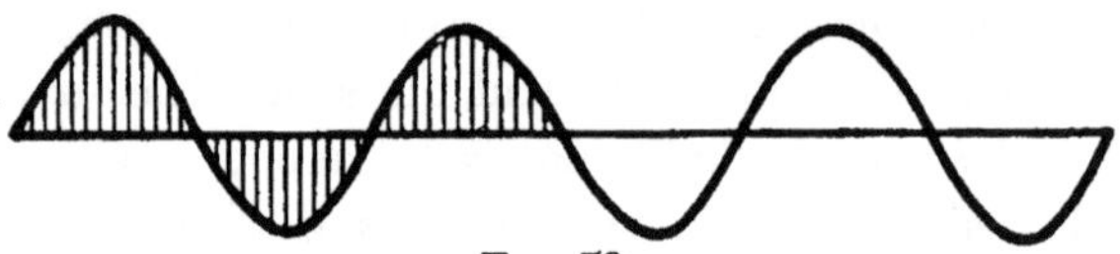

FIG. 79

"And so with the area in 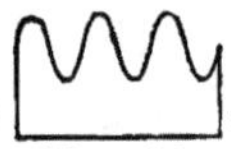 . The area becomes a rectangle.

FIG. 80

Even if the curve is shifted!

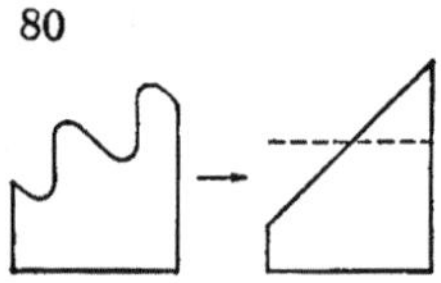

FIG. 81

This is a question of symmetry and equilibrium within the whole figure.

"What of other curves? Of course it is true also for $y=x$ (see Fig.

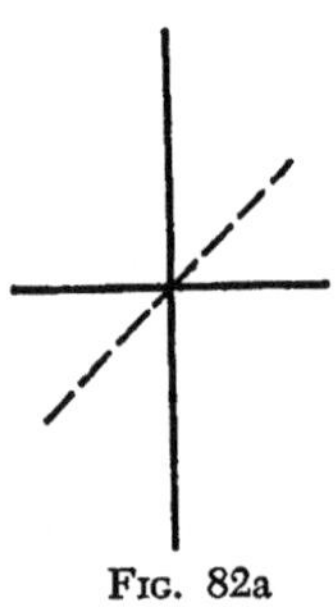

FIG. 82a

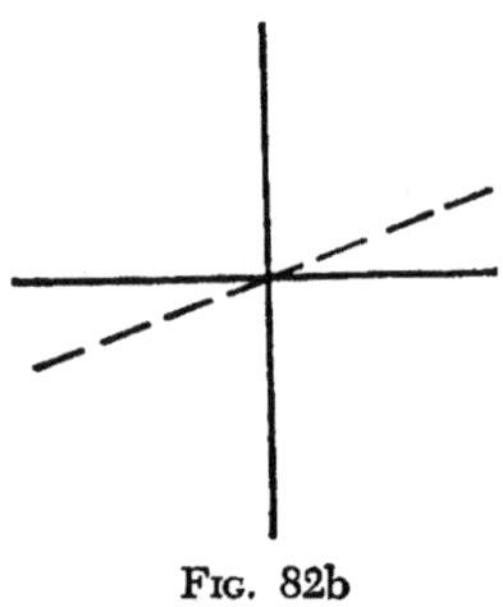

FIG. 82b

82a) or $y=ax$ (see Fig. 82b). It holds, however the angle may change, with any straight line symmetrically cut. In $y=ax+b$ the line is only shifted. And the area is the height of the center times the base in figures such as the following:

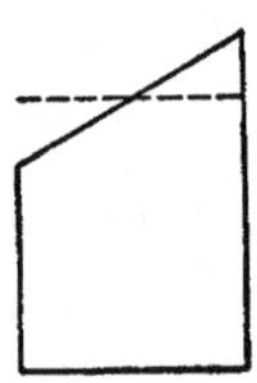

FIG. 83

"It holds with corresponding series $x_n = x_{n-1} + k$. The sum of the terms is the value of the center times the number of digits, c times n."

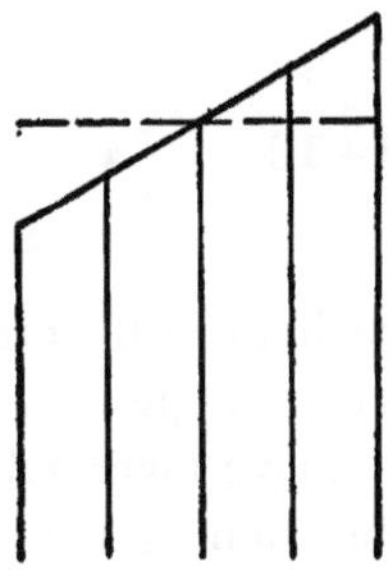

FIG. 84

In this way he arrived at the Gauss theorem, not from a series starting with 1, but from the view of the structural whole-quality of the balance in the distribution as seen from above.

I shall presently return to his thought process. In the meantime it will be realized that this is not just a question of stating the individual differences between successive terms, of stating the equality of such differences, etc., or of dealing with class laws of series—laws which somehow hold for all the points or items. It is fundamentally a question of balance within the whole, of realizing equilibrium with regard to whole-features. And this balance is a dynamic thing, sensitive to deviations or disturbances in any of the parts.

If one makes a diagram of the points of such a Gauss series, one sees that the line is straight, or that there is a deviation from straightness, a disturbance, long before one can state or know the measure of the differences, and whether they are equal, etc. For example,

1+2+3+4+6+7+8

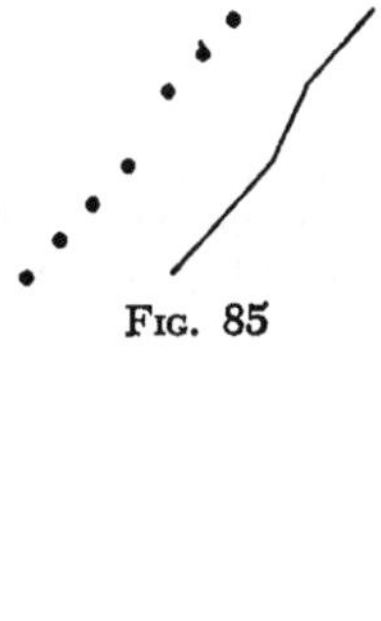

FIG. 85

or

1+2+3+4+6+8+10

FIG. 86

There is sensitivity to such disturbances which stand out against the clear whole-quality of straightness. Such series, for example, the first of the two series above (without the 5), can be formulated as lawful series in a consistent general formula, $x_n = f(x_{n-1})$, no less lawful than the straight series, only more complicated in its terms. But the series $x_n = x_{n-1} + k$ is outstanding in its structural simplicity, in the structural clearness of its whole-quality. Nobody would spontaneously view the series

$$1+2+3+4+5+6+7+8,$$

especially in its diagram, as a deviation from the more complex formula, with the 5 intruding as a disturbance, even though in the usual mathematical terms a law is a law, each like every other.[23]

Similarly with the sine curve, or with dots in a sine curve. Long before one states or knows the measure of the distances between the single points, long before one has a "class law" for them, one sees—viewing the whole—that the path of the curve is regular.

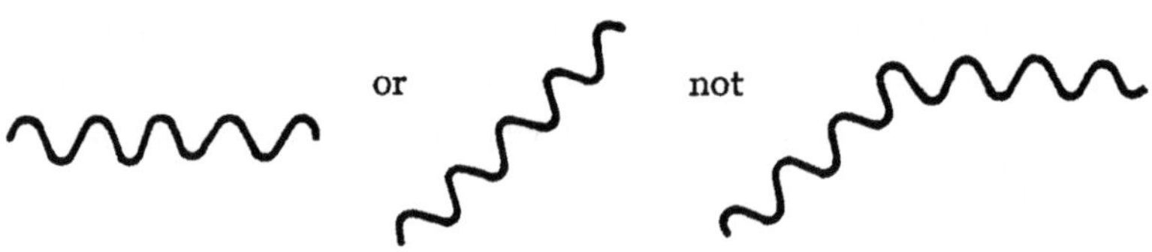

FIG. 87

[23] Of course the facts have to decide. One may be mistaken in making the structurally simpler assumption. What is decisive is the structural behavior of the terms in the series. Cf. p. 116, footnote 22.

One sees that there is rhythmical repetition in regular sub-wholes,

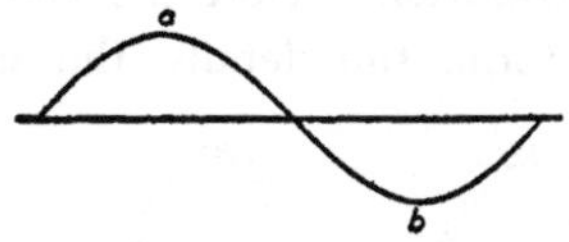

that b corresponds to a;

FIG. 88

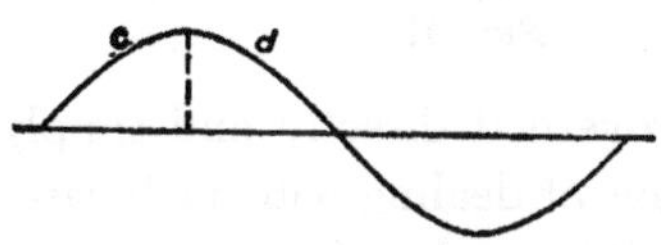

that c corresponds to d.

FIG. 89

One grasps symmetrical features in the sub-wholes from above. What first matters here, psychologically, are the outstanding features of the whole,[24] and of the sub-wholes. From these as centers, deviations often become conspicuous, are conceived *as* deviations.

Many will say, "Quite so; but these are only inexact, global, psychological views, not to be compared with exact mathematical formulations in terms of $y=f(x)$, etc." The objection is blind. Is the mathematical way, the exact way, necessarily the way from below? From the elements? Must it, to be exact, derive whole-qualities such as symmetry merely secondarily? Should there not also be mathematical ways from above, no less exact? Mathematical ways that start from whole-qualities and go secondarily from these to the elements?

Psychologically, the situation with regard to viewing whole-qualities is frequently not essentially changed if, instead of a sine curve exact in all its details, one looks at a crinkled sine curve,

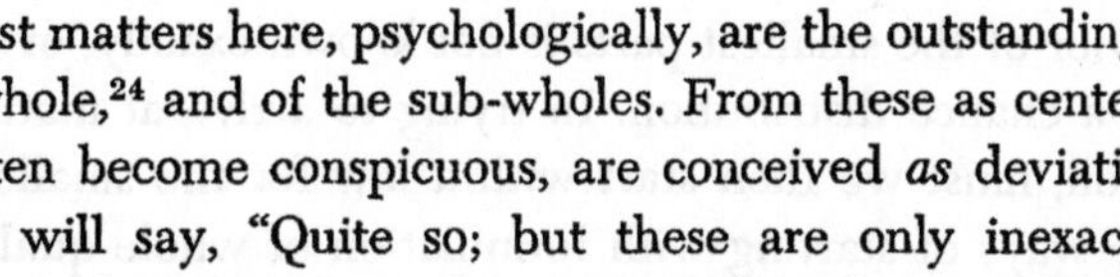

FIG. 90

[24] This holds not only for rhythmical forms and symmetrical features; it holds also for the direction of a main vector in changes, trends, etc.

It holds for thinking processes and for action if, in spite of complications, in spite of small deviations, one does not lose the clear whole-direction.

or a curve drawn in dots in which there is some distribution of errors, even a chance distribution.[25] Here the view from above gets the whole-qualities, the form; the details, the smallest parts, the

FIG. 91

elements are no longer governed directly and simply by the law. Mathematics has exact ways of dealing with such cases, formulating whole-qualities that are valid in spite of variations in the parts.

This is often the situation in modern physics. In such cases we know about whole-qualities, the whole-behavior of systems of which the behavior of the smallest parts is not known exactly, or is known to follow a chance distribution. In trying to arrive at mathematical formulation, must we then start with a law for the smallest parts? There are ways of starting with formulation of whole-qualities that allow for varying behavior in the smallest parts.

Should it not be possible, moreover, to develop methods for dealing with questions of dynamics in this way? To conceive of tendencies toward certain transformations not on the basis of the and-sum of elementary piecemeal forces but as functions of whole-qualities and their disturbance?

Whatever future developments may be, it is certainly not true that the way from above is merely a "global," "inexact" way; what is true is only that the techniques of the other way are mathematically more developed.

Let us now return to the process reported on p. 115 ff. Although similar in its view of the Gauss problem to the procedure of some of the other subjects (cf. §II), there is still quite a difference. This subject took the problem in a broader, deeper way. To him it

[25] At the International Congress of Psychology in Groningen, 1926, I reported on investigations in this regard in the context of a paper dealing with gestalt problems on perceptual thresholds ("Zum Problem der Schwelle," *Bericht über den VIII Internationalen Kongress für Psychologie*, Groningen, 1926). And some years ago Woodworth mentioned an illuminating example: a prehistoric ringwall was discovered from an airplane in a field which had been cultivated for many decades. Nobody had noticed it before. Nobody had been able to see it without the broad survey of the whole field which the pilot had.

offered not merely the nice possibility of reorganization in the particular task; no, he was focussing on what opened up in the inner relation between form and sum.

In his next step he compared his formula $c{\cdot}n$ with the Gauss formula $(n+1)\frac{n}{2}$ and remarked that it became $c{\cdot}n$ with the slight change into $\frac{n+1}{2}{\cdot}n$. Then he said, "$c{\cdot}n$ seems to be the more general, more fundamental formula in which it does not matter what the amount of the equal differences is, what the end term or the first term is. On the other hand, I have two variables here, c and n, whereas for the Gauss formula one variable suffices, n! So the Gauss formula is the more elegant, requiring only one variable." Suddenly he laughed and said: "No, it is only an external limitation that makes it appear so. This '1' in $(n+1)$ is in fact the first number in the series just as n is the last number. Similarly $\frac{n}{2}$ looks only superficially so simple. It is in fact $\frac{z-a+1}{2}$, z being the last, a the first number in the series. It turns into $\frac{z}{2}$ only because $a=1$. That the series should start with just the value 1 is of no importance. It is a special case. Furthermore, that particular kind of formula is a special case because it is limited to a difference of 1 between the terms. The important thing is the basic relationship: some series, some curves, some distributions show this clear inner relation between their whole-qualities, their principle of construction, and their sum. I want to find out more about this. What are the requirements generally? At bottom it seems to be a question of balance in the whole, of compensation at some level among the various parts." Thinking then of the matter of compensation, he recognized that the same principle must hold for products. Although he immersed himself in these problems, I shall not follow his later steps here. They led him to the question of whether an inner relation between a growing series and its sum is possible only by way of compensation, and eventually to the facts of finite limits of infinite series.

In such processes of thinking the solution of an actual task,

"Problem solved, task finished," is not the end. The way of solution, its fundamental features, the problem with its solution function as parts of a large expanding realm. Here the function of thinking is not just solving an actual problem, but discovering, envisaging, going into deeper questions. Often in great discoveries the most important thing is that a certain question is found. Envisaging, putting the productive question is often more important, often a greater achievement than solution of a set question, just as here in our example the main thing seemed to be to envisage, to crystallize a basic structural problem—a much broader, much deeper process than the processes reported before.

Just as a task, a problem situation in productive thinking, is not something closed within itself, but tends toward its solution, its structural completion, so even a task with its solution is often not a thing by itself. It again may function as a part that points beyond itself, striving to envisage, to clarify a broader field. Often such a process takes a long time; it is a drama with setbacks and struggles. There are fine cases in which the process proceeds irresistibly, through months, through years,[26] never losing sight of the deeper issue, never getting lost in petty details, in detours, bypaths.

This is one important difference between pedantic thinking and thinking in broad lines, a difference that in life as well is terribly important. Many theorists do not rightly see this difference or its importance, they confuse it with questions of exactness, of piecemeal one-sided exactness which is blind to the great features at issue. But exactness is not inconsistent with great features; it is a needed co-operator.

[26] This is true not only of individuals but of groups, so that great problems are handed on from generation to generation, the individual acting not primarily as an individual but as a member of the human group.

CHAPTER IV

Two Boys Play Badminton; A Girl Describes Her Office

IN THE previous chapters one issue played a particularly important role: the factor of reasonable reorganization, reorientation, which enables the subject to view[1] the given situation in a new[2] and more penetrating[3] perspective. It is this factor that leads to or constitutes

[1] Dislike of terms like "viewing," on the part of behaviorists and operationalists, should not blind them to the problem. Such terms are, I think, quite proper. But the main issue can also be formulated in the terms of these extremists, and it still remains essentially the same. Even if, strangely enough, one should want to disregard facts of conscious experience entirely, the consequences of reorganization show up in changes of objective behavior. In a first investigation, what really matters in the term "view" can be exactly formulated in terms that are operationally defined.

[2] For instance (Chap. I), the transition from 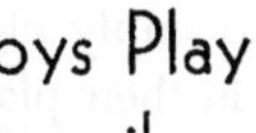 into 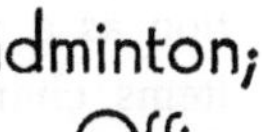
from  into 
e.g. (Chap. III), the transition from ⟶ of the Gauss series as first seen to the new view ⟶⟵
also below (Chap. V), the transition from the sum of the exterior and interior angles of a polygon to the sum of the angles δ (see Fig. 129) plus two right angles, from the exterior angle 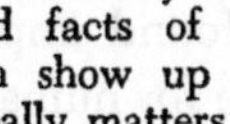 to 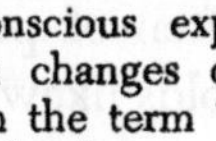
from the interior angle 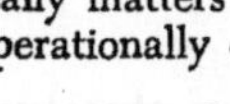 to 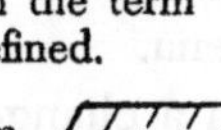
again, similar reorientations in the chapters on Galileo and Einstein.

[3] For instance (Chap. V), the way in which the sum of the angles in a closed figure or in a solid is conceived; the thought process last reported in Chap. III; pp. 115-123; also the emergence of a more penetrating view in the chapters on Galileo (Chap. VI) and Einstein (Chap. VII).

a discovery in a deeper sense. In such cases a discovery does not merely mean that a result is reached which was not known before, that a question is somehow answered, but rather that a situation is grasped in a new and deeper fashion—whereupon the field broadens and larger possibilities come into sight. These changes of the situation as a whole imply changes in the structural meaning of part items, changes in their place, role and function, which often lead to important consequences.[4]

Before the thought process takes place, or in its early stages, one often has a certain whole-view of the situation, and so of its parts, which is somehow unsuited to the problem, superficial, or one-sided.[5] Such a first inadequate view often prevents a solution, a proper dealing with the task. If one sticks to this view, one will often be unable to solve the actual problem. On the other hand, when the change has occurred, and the problem has thereby been solved, one is sometimes astonished to see how blind one has been, how superficially one had viewed the situation.

The change in structural view according to the requirements of the problem is often of profound importance in the development of science. This holds likewise for human life in general, particularly in social phenomena.

Of course, such a change of structure is to the point only if the proper view has not been there from the start. Often the first view has only lacked adequate penetration, sufficient clarity; or some demand implied in the situation may not have been fully realized. In such instances the solution requires mainly that the situation be further clarified or crystallized, and that aspects or factors be realized which were only vaguely present in the first view.

In order to study such transitions and their consequences for the role and function of parts, I have sometimes employed specific experimental sets which bring about a sudden radical transforma-

[4] For instance, in Chap. III, p. 115, "+1" becomes the zero of the "real series," the "0" becomes "−1," etc.; also in the same chapter, p. 90, the 7, conceived first as 6+1, changes into 8−1.

[5] Cf. the examples in M. Wertheimer, "Über Schlussprozesse im produktiven Denken," *Drei Abhandlungen zur Gestalttheorie* (Erlangen, 1925) pp. 164-184; W. D. Ellis, *A Source Book of Gestalt Psychology* (Harcourt, Brace and Company, 1939) selection 23.

tion from a first to a second view. Some simple examples have been mentioned (cf. Chap. I, pp. 48-49 ff.).[6] Often powerful, dramatic occurrences take place when a subject experiences the change. The same sets also permit a study of what happens to the various regions of the structure when it changes: organization and grouping of the parts; changes in the location of "caesuras"; centering; shifts as to structural relevance; the appearance of gaps, of disturbances; changes as to the extent to which local conditions may be varied; changed trends in the subject's expectations, in the emergence of new whole-tendencies, and in the direction of requirements.

When such transitions occur in thinking, it is not the ease of arbitrary change as such which characterizes intelligent behavior; nor is it the ability in a given situation to produce one or another or a third structural view at will, as some seem to believe. Rather what matters here, and what is characteristic of intelligent processes, is the firm transition from a less adequate, a less proper structural view to a more sensible one. And in fact experiences seem to show that sensible persons, real thinkers who are often quite capable of making sensible transformations, also children, show little ability and even less willingness to engage in *senseless* changes of given situations.

Sometimes the transition from an unstructured and-sum to the appropriate structure is needed. But even more significant is the transition from a one-sided view, a superficial structurization or misstructurization, from a miscentered, distorted or poor view to the adequate and properly centered structure.

The main cause of unreasonable, blind behavior seems to be that a person sticks to an old view from sheer perseveration and habit, which make him ignore or even actively deny the more reasonable requirements that are clearly indicated in the situation.

In order to show more clearly how such transitions come about, I will now report on some simple examples from everyday life which I studied in various experiments.

[6] Cf. also M. Wertheimer, "Zu dem Problem der Unterscheidung von Einzelinhalt und Teil," *Zeitschrift für Psychologie*, 1933, Vol. 129, pp. 353-357 and a description of further examples from my lectures in M. Scheerer, *Die Lehre von der Gestalt* (Walter de Gruyter & Co., 1931) pp. 209-210.

I

Two boys were playing badminton in the garden. I could hear as well as see them from my window, although they did not see me. One boy was twelve, the other ten years old. They played several sets. The younger was by far poorer, he was being beaten in all the games.

I heard some of their conversation. The loser—let us call him *B*—became more and more unhappy. He had no chance. *A* often served him so cleverly that he could not possibly return the bird. The situation grew worse. Finally *B* threw down his racket, sat on a tree trunk, and said, "I won't play any more." *A* tried to persuade him to continue. No answer from *B*. *A* sat down beside him. Both looked unhappy.

I interrupt the story here to put a question to the reader: "What do you suggest? What would you do if you were the older boy? Do you have a productive proposal?"

If in telling the story one stops at this point, some subjects clearly do some thinking. They make remarks which show that they see a considerable problem, and they venture some suggestion as to what the older boy should do.

Most subjects do not. They are dissatisfied with the interruption of the story, wait to have it continued, are puzzled because I do not go on, ask how things actually developed, what the boys did, whether they were my boys, why I told the story, why I stopped there, whether this was an experiment—whether later I would perhaps give a memory test, and so forth.[7]

Some reminisce, reflect, meditate, "Oh, such situations are very familiar to me. You know I am interested in children. This reminds me of my uncle's trouble with his two children." Or they recall a paragraph from a textbook on child psychology.

Some make nice subsumptions under general topics: "This is a case of . . ." and classify the case, then often embark on generalities, more or less needlessly, as, for instance, about social adjustment, adaptive behavior, problem children.

[7] The extreme case, of course, is when nothing at all happens. Finished? Finished. Are you going to tell more stories? What will we do now?

Some want to know more of the facts, ask a number of questions, more or less reasonable ones.[8] For instance, "Of course the older boy had had more practice?" Or: "Was the younger boy generally slower in his reactions?" Even psychoanalytic questions are asked.

For the most part these subjects do not produce any concrete, productive proposals and show some surprise if asked for one directly. It is clear that they have not thought of that possibility at all, busy as they are with recollections, classifications, or fact-gathering.

If asked directly, nearly all do suggest something. Often in the tone of "What should be done in such a situation is obvious." In most cases the answers are clearly given without any effort of thought, as sheer reproductions of things seen or heard before, as applications of some known rule of behavior, sometimes remembered from courses in educational psychology. Often they are given in a low tone of deep conviction, often with a superior air.

The proposals offered are frequently characteristic of prevailing ideas about children, about human beings in general, about morals, current social rules, and doctrines to which the subjects adhere. Often one can recognize doctrines of various schools of psychology in their far-reaching consequences.

The usual types of suggestion were as follows:

"Promise the younger boy a piece of chocolate."

"Start another kind of game, say chess, in which the younger is equal or even better. Or promise to play, alternately with badminton, some game in which he is by far the better."

"Bring him to reason by plain scolding. He should be manly, not a sissy. Can't he take it on the chin? He has to learn to take it. Use superior authority in bringing the younger boy to reason."

"Don't bother with him, he is a sissy. It will be a lesson to him."

"Offer a handicap."

"Promise the younger boy that the older boy will not make full use of his superior power and skill."

The reader may later compare these proposals with the boys'

[8] The extreme case was a nice young man who in this case—and in fact always—started at once asking question after question, lots of them. You could not stop him. Curiosity is not always, not in every sense of the word, a sign of intelligent thinking or of reasonable behavior.

own solution, not merely with regard to fitness—some of these proposals are proper, since they depend on actual conditions in the real situation—but with regard to the kind of thinking that is involved.[9]

Sometimes, as I have said, proposals are not made in a facile, casual way, by easy recall or application, but after serious thinking with a sense that the problem touches deep questions. There is earnest deliberation, productive questions are asked. Several times the thought processes contained steps similar to those which occurred in the boys.

I shall now continue with the story. In addition I shall try to describe what I think was going on in the boys' minds.

1. "I am sorry. Why don't you go ahead?" said the older boy in a sharp, angry voice. "Why do you break up the game? Do you think it's nice, to stop in this silly way?" He wanted to continue. The refusal of *B* made it impossible. He liked to play, he liked to win; it was even nice to be able cleverly to trick one's opponent with the service. *B* is the disturber, he makes it impossible for *A* to do what he so much wants.

2. But it was not so simple. At the same time *A* was not quite at ease, he did not feel nice about it. After a while, during which curious things were going on in his face—I wish you could have seen him as he frequently glanced sidewise at *B*, and away again—he said, but now in quite another tone, "I am sorry." Obviously a radical change had taken place—clearly *A* now felt sorry that the other boy was so unhappy. He had realized what was going on in *B*, how the situation looked to the other fellow.

Perhaps a sad, quiet look of *B*'s had helped, as *B* once turned his head toward *A* for a brief moment. *A* realized—it was not a quick process, it took some time—why the younger boy was sad, why

[9] Also with regard to the philosophy of life implied, and the underlying psychological doctrines which often come out in discussion: e.g., the naïve pain-pleasure principle, reward and punishment psychology, submission to the idea that consent may be bought, horse-trading in an extreme form ("You be my slave now, then I'll be yours"); on the other hand, a moral appeal which often works productively but which under certain circumstances may degenerate into sugar-coating the pill.

Often these suggestions go with a kind of cynicism; or they are treated with easy superficiality as obvious, properly applied psychology.

with no hope of holding his own he felt like a victim. For the first time *A* felt that his way of playing, his shrewd service, looked to *B* like a mean trick; that *B* felt he was not being treated fairly, that *A*'s actions were unfriendly to him. And *A* felt that *B* was somehow right . . .

Now he also saw himself in another light. To serve as he had done, without giving *B* the slightest chance to return, was something more, something other than skill.

3. "Look here," he said suddenly, "such playing *is* nonsense." It was now nonsense not only for *B*, but also nonsense for him, nonsense for the game itself. Thus the trouble had become of a deeper kind.

Expressed in adult language it was as if he had thought—he certainly did not, he only felt his way through—"It is senseless for two to play together in such a way. The game requires some reciprocity. Such rank inequality does not suit the game. The game becomes real only if both have some hope of achievement. The game changes its whole character, becomes a bad thing for the one player, for the other, for both, if there is no such reciprocity; without it there is no longer really a game—a tyrant merely makes his victim run around."

4. Then the expression of his face changed. He looked like someone trying with difficulty to grasp something, on whom something slowly begins to dawn, and he said: "Such a game is a queer thing. I am not really unfriendly to you . . ." A vague idea had arisen in him of what a grownup would call the "ambivalence" involved in the game: on the one hand, it is so nice to have a good time together, to be good friends; on the other hand, there is this effort to overcome your opponent, to beat him, to make his achievement impossible, which in certain circumstances may be felt as, or may actually become, plain enmity.

5. Thereupon a bold, free, and deeply consistent step occurred. He mumbled something like "Must it . . . ?" Clearly he desired to face the trouble squarely, to deal with it honestly, directly. I interpret this "Must it?" as "Is this factor of enmity necessary when it spoils all that is nice in the game?" Here the pragmatic problem arose: "How can I change it? Is it not possible to have, not one

against the other, but—" His face lit up and he said: "I have an idea—let us now play this way: let's see how long we can keep the bird going between us and count how many times it goes back and forth without falling. What score could we make? Do you think we could make it ten or twenty? We'll start with easy serves but then let us make them harder and harder . . ."

He spoke happily, like someone who had discovered something. It was a new thing to him, likewise to *B*.

B agreed happily: "That's a fine idea. Let's." And they began to play. The character of the game was entirely changed; they were co-operating, acting together, working hard and joyfully. *A* no longer showed the slightest tendency to trick *B*; to be sure, his moves became more difficult, but he called out, "A harder one, can you get that?" in a sensible, friendly way.

Several days later I saw them play again. *B*'s playing was greatly improved. It was a real game. As one could see in his later behavior, it was quite an experience for *A*. He had discovered something, gained something that went far beyond the solution of a little problem in a game of badminton.

The solution in itself may not amount to much so long as it is seen merely from the outside. I do not know whether badminton or tennis experts would agree that it was appropriate.[10] It does not matter. This solution was not just a technical affair for this boy. It involved the transition from a superficial attempt to get rid of a trouble to facing the fundamental structural issue and to dealing with it in a productive way.[11]

What were the steps that led to this solution? Of course, when

[10] I know chess players who are intensely disturbed if fine constellations in a game are disrupted by some peripheral, thoughtless mistake. They hate it equally, no matter who makes the mistake. Some have the custom—*horribile dictu* for experts—of correcting such errors. Why? They like a good game; they have no interest in winning by some silly mistake of their opponents. Sometimes they even co-operate in order to make the game more perfect.

In general I think there should be more games of the co-operative, creative sort instead of merely competitive ones.

[11] This would by no means have been a reasonable proposal under all circumstances. If A had been a ruffian, seeking only his own advantage, bent upon conquering, whatever it might mean to the other fellow, proud even of the pain he inflicted on him, and if *B* then would have proposed it—it would have been strictly senseless behavior. In fact, the opposite would have been called for: to quit entirely or to train somewhere else for the real fight.

one interprets a single case, the factual basis, the basis for conclusions, is small. Still, let us try to formulate the essential points.

Initially *A* saw the whole situation, and in it *B*, the game, the trouble, centered by his ego. They were determined, by this center, in their meaning, role, place, function both as to thinking and as to action. If this were an extreme case, *B* would be nothing but someone whom *A* needs in order to win; consequently, in refusing to play, *B* would be the "disturber." The game would be "the thing in which I display my abilities, in which I win." *B* represents a barrier which is in the way of *A*'s egocentric requirements, vectors, acts.

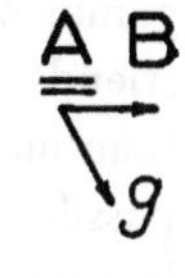

FIG. 92

A did not persist in this one-sided, superficial view. He began to realize how the situation looked to *B*; to *B* as the center. In this recentered structure he now saw himself as a part, as a player who dealt in a not altogether nice way with the other player.

FIG. 93

A bit later the *game* itself, its whole-qualities and requirements, became the center. Neither he nor the other was now the center, both were viewed with reference to the game.

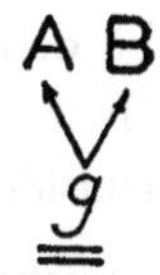

FIG. 94

Logically, *A* (as *A* views himself) differs in the three centerings;[12]

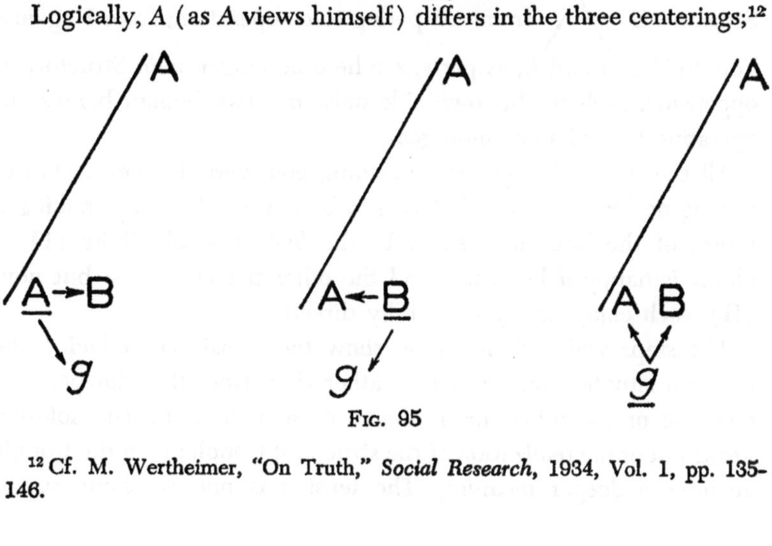

FIG. 95

[12] Cf. M. Wertheimer, "On Truth," *Social Research*, 1934, Vol. 1, pp. 135-146.

so do the other items, along with the dynamic requirements, the vectors, the actual situation. The game is now clearly seen as a deviation from the "good game."

But what in the structure of the game itself lies at the root of the trouble? There is a delicate functional equilibrium in the good game: on the one hand, having a nice time together, behaving as friends; on the other hand, "trying to beat him." Attitudes deeper than mere external rules of fair play make this delicate equilibrium possible, establish the differences between a good game and a ruthless fight or a ruthless competition; in short, make a game a game. The equilibrium is psychologically delicate; it may disappear—as it did in this situation.

The "against," the "trying to beat him," that functions properly in the good game had grown into an ugly feature which no longer fitted the game situation. That gave rise to the vector: "What could be done about it? And done immediately?" Here lies the trouble. "Is it not possible to get at the core of the situation?" This leads to considering structure II.

Structure Ia ——→ ←—— or, with these boys, rather

Structure Ib ———→ ←— is transformed into

Structure II, Λ from competing to co-operation; from I against You, to We. *A* and *B*, as parts, are here no longer as in Structure I, opponents, each for his own side only, but two human beings co-operating toward a common goal.

All the items change their meaning considerably. For instance, serving no longer means that *B* is to be overwhelmed by making a return of the bird impossible. In the first state of affairs (I), a player is happy if he scores and the other player misses; but now (II), each enjoys every good rally directly.

The steps which then follow show the transition to facing the problem situation on *its* merits rather than from the viewpoint of one side or the other, or of the mere sum of both. The solution grows out of the realization of the structural trouble; then the trouble acquires a deeper meaning. The tension is not overcome superficially; rather, the new direction of the vectors follows from basic structural requirements which demand a genuinely good situation. You may think I read far too much into the boy's mind. I don't think so. You may have far too poor an idea of what can happen in the minds of boys.

In short we note specifically:

1) Operations of recentering: transition from a one-sided view to the centering required by the objective structure of the situation;

2) a change of meaning of the parts—and of the vectors—in accordance with their structural place, role and function;

3) a view of the situation in terms of "good structure," so that everything fits the structural requirements;

4) a trend to go straight to fundamentals, to face the issue honestly, and to draw the consequences.

I want to remark that the feature of straightness, honesty, sincerity, does not seem peripheral in such a process. Generally speaking, it is an artificial and narrow view which conceives of thinking as only an intellectual operation, and separates it entirely from questions of human attitude, feeling, and emotion—"because such topics belong to other chapters of psychology." This is especially clear in one example, in the transition from a blind egocentric view with its emotional ingredients to the later steps. But even seemingly mere intellectual processes involve a human attitude—that kind of willingness to face issues, to deal with them frankly, honestly, and sincerely. Although I have referred to this fact only briefly in other chapters, it seems essential in many cases of productive thinking, including even our problems in pure elementary geometry.

Of course it may happen that certain solutions are found when the subject assumes just the opposite attitude, when he tries to subdue the situation by a trick or to do violence to it. But here too such factors seem not merely to accompany the "intellectual operations"; rather, the very nature of the operations, their genesis, and their development seem deeply connected with the actual human attitude toward the problem and its solution. Neither slyness, nor a spirit of domineering seems to be the most advisable attitude in productive thinking, even though it may sometimes lead to practical success, and thus make for a certain quick, short-range efficiency.

Connected with these matters is a further point which seems of fundamental importance for a true understanding of productive thinking both in the theoretical and in the practical realm. This is the transition from a first stage in which the subject simply wants to reach a certain goal—in which his thought is entirely centered by this goal—to another stage in which the vectors, the operations, the actions become centered by more fundamental requirements of the situation. In a sense a subject may become virtually blind if he looks only at that goal, and is entirely governed by the urge toward *it*. Often he must first forget what he happens to wish before he can become susceptible to what the situation itself requires. And so, in better instances, his attitude will be far more similar to the attitude of an obstetrician or a wise adviser than to that of a clever and domineering conqueror or aggressor.

This transition is one of the great moments in many genuine thought processes. The role of the merely subjective interests of the self is, I think, much overestimated in human actions. Real thinkers forget about themselves in thinking. The main vectors in genuine thought often do not refer to the I with its personal interests; rather, they represent the structural requirements of the given situation.[13] Or when such vectors do refer to an I, this is not just the I as center of subjective striving.

Of course the transition may lie also in the direction of the deeper requirements of the I itself. Sometimes there is a happy coincidence between the requirements of the situation which presents the problem and the real, the deeper needs of the I, as was the case with our boy.

This was only a modest little story. Yet some of its characteristics, as, for instance, the essential progress which occurred through a change of centering, are, I think, characteristic of profoundly important accomplishments[14] in human beings, in human society. One can sometimes observe marvelous changes in individuals, as when some passionately biased person becomes a member of a jury, or arbitrator, or judge, and when his actions then show the fine transi-

[13] See E. Levy, "Some Aspects of the Schizophrenic Formal Disturbance of Thought," *Psychiatry*, 1943, Vol. 6, pp. 55-69.

[14] M. Wertheimer, "A story of three days." In R. N. Anshen (ed.), *Freedom; its Meaning* (Harcourt, Brace & Co., 1940) pp. 555-569.

tion from bias to an honest effort to deal with the problems at issue in a just and objective fashion. The very development of the idea and institution of courts of justice is at issue here.

Centering—the way one views the parts, the items in a situation, their meaning and role as determined in regard to a center, a core, or radix—is a most powerful factor in thinking. The problems of centering have been neglected in traditional logic and in psychology. Strong forces are at work in centering, in envisaging—or in trying to envisage—the true center according to the nature of a situation; but they are equally strong in cases of blind, enforced or willful miscentering, so efficiently used in some kinds of political propaganda. Although there are many strong forces working against true centering, there is nevertheless in human beings a clear desire not to be structurally blind, a desire to center properly, to do justice to the situation, to center in accordance with the nature of the object, of the structural objective requirements.

As to the concept of centering, it seems to be tacitly recognized that *adequate* centering, with its consequences for objectivity and justice, is of the utmost importance. Otherwise, why should actual miscentering be the more energetically disguised as true and just centering, the less it is so in fact?

II

Quite a number of things are involved in the process of recentering as it occurred in the badminton example. Some basic points will become clearer if we now consider a simpler story.

I was visiting a family. The daughter of the house came home and was introduced to me. Her father asked how she had spent the day. She answered that there had been a good deal of work, but that she was fine. I asked, "Are you working?" "Yes," she answered, "I am working with a firm." "Is it a big place?" "Well," she said, "there are a number of people in the office. I have to do directly with a Mr. *A*, a Mr. *B*, and a Mr. *C*, who often come to my desk, ask questions, bring me letters, etc. There are others in the office with whom I do not have to do directly. Mr. *A* has dealings with a Mr. *D*, Mr. *B* with a Mr. *E*, and Mr. *C* with a Mr. *F*. *D* and *E* also have dealings with each other; so do *E* and *F*. Let me see, that makes six people in the office besides myself."

I asked, "Are you the boss?" "Oh no," she answered. "Do you give orders to anyone?" "Oh yes, I sometimes give orders to Mr. *A* and to Mr. *C*. I get orders from Mr. *B*; Mr. *D* gets them from Mr. *E*, Mr. *E* from Mr. *B*, and Mr. *F* from Mr. *E*." (She was apparently logically minded, and was trying to tell the whole story.)

I was somewhat disturbed—I guess the reader is too—and I said: "I am still in the dark about the people in your office." "Oh, but I have told you everything," she answered. Nevertheless, I was in the dark. Suddenly I said—it was a hunch—"So Mr. *B* is your boss, and you are directly below him, and so is Mr. *E*?" "Yes," she said.

As she saw her office, she had stated the relations correctly and completely, and yet she had not made the real picture clear. Most people, if asked such a question, would start with some such statement as "I am working directly under the Boss *B*; so is a Mr. *E*"; they might add, "Mr. *E* and I each have two people working under us to whom we give orders," and might continue, "Two of these sometimes have business with each other." This would have been a sensible description; it would have given a clear picture of the structure of the office setup. But this girl had enumerated the men and the relations in a bewildering succession, in fact in a way that was blind to the structure of the situation; she had centered everything about her ego—except for the last bewildering statement in the second description which did not refer to her.

This is an innocent example of a foolish attitude in life and in thinking, which often has considerable consequences for shaping one's views and one's actions.

Of course it might have been mere clumsiness in description, but from the girl's subsequent remarks and behavior it became clear that this was characteristic of her real attitudes. Some time later, when I met one of her coworkers from the office, I asked him how she was getting on. "Quite nicely," he said; "she is a fine person. But we are not certain whether her job will last very long. She has a funny way of behaving toward the others and even toward her work. She seems to relate all things to herself, as if she were always the center of the situation, even in business matters when nobody is thinking of her personally. That is not good for business."

When self-centering is extreme, it becomes a well-known symptom of a psychopathological state that often leads to a precarious situation both in social and in personal affairs.[15] Self-centering is not at all the general, the natural attitude, as some influential views of our time would have us believe.

Let us look more closely at what this girl did in her description. In a schema it may be represented as follows:

First Description

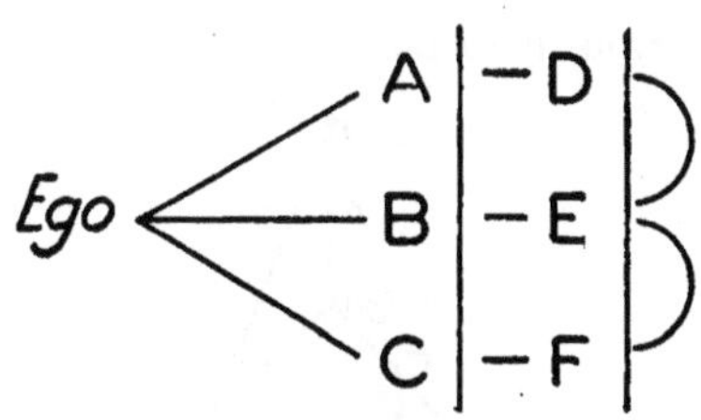

Fig. 96

This schema is quite similar to the way a logistician would write a list of relations in a relational network. He would state the relations *Ego r X*, and so forth, as follows:

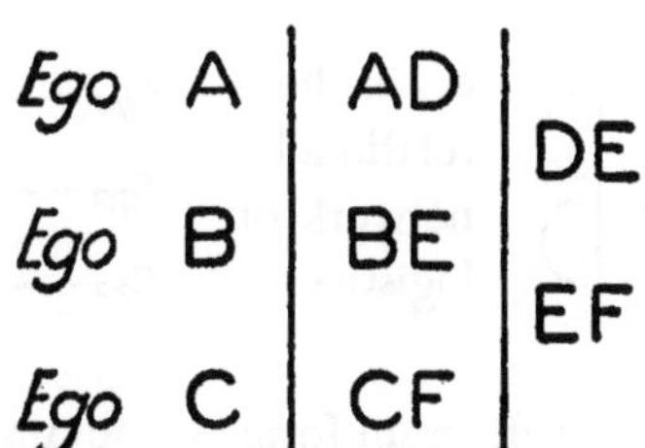

Fig. 97

If one asks a person for a diagram that would fit the girl's description, a picture like this is usually given:

[15] Cf. M. Wertheimer, "Über Gestalttheorie," (Erlangen, 1925). (Editors' note: this article is translated into English under the title "Gestalt Theory," *Social Research*, 1944, Vol. 11, pp. 78-99); also W. D. Ellis, *op. cit.*, selection I; E. Levy, *op. cit.*, pp. 59-69; H. Schulte, "Versuch einer Theorie der paranoischen Eigenbeziehung und Wahnbildung," *Psychologische Forschung*, 1924, Vol. 5, pp. 1-23.

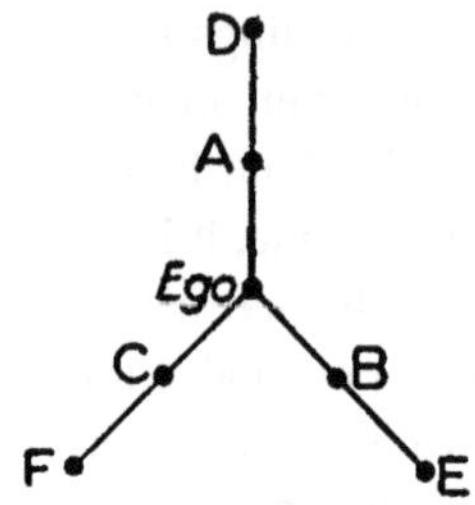

Fig. 98

which clearly shows the ego as the center. This characteristic is usually not changed when the last two lines from *D* to *E* and from *E* to *F* are added, although some people will ask, "Is there not also a connecting line from *F* to *D*?"

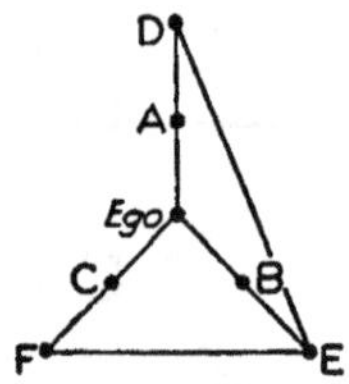

Fig. 99

Second Description

In this description the girl qualified the relations by giving them directions. But the list now constitutes a veritable tangle of such directions:

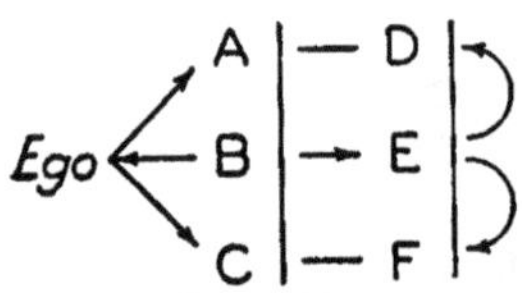

Fig. 100a

or, in the relational network of logistics:

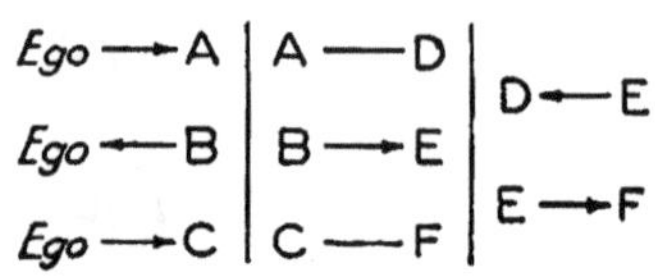

Fig. 100b

If we represent this in diagram form we obtain:

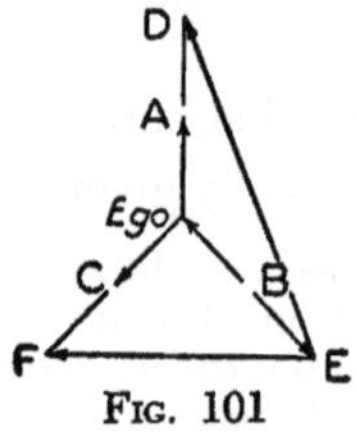

Fig. 101

At this point something usually happens to those who look at the diagram. It begins to appear as though "wrongly drawn," "distorted." It may then be changed into a form which we will give in a moment.

Third Description

Both a schema and a diagram of the adequate description look fundamentally different:

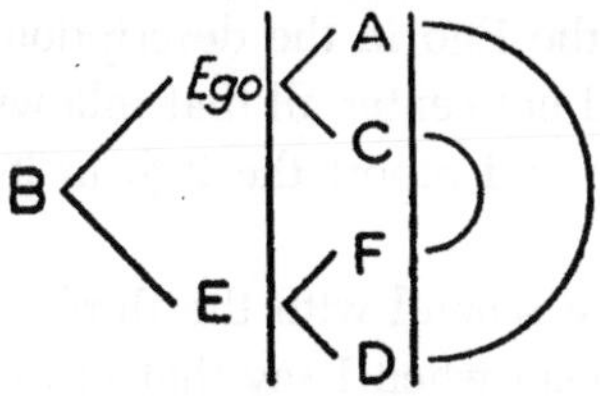

FIG. 102

or, in the relational network

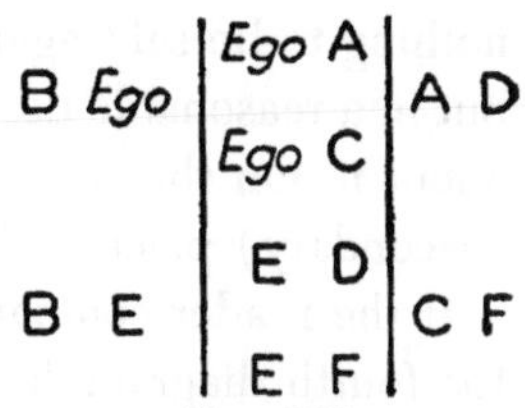

FIG. 103

As a diagram one gets here:

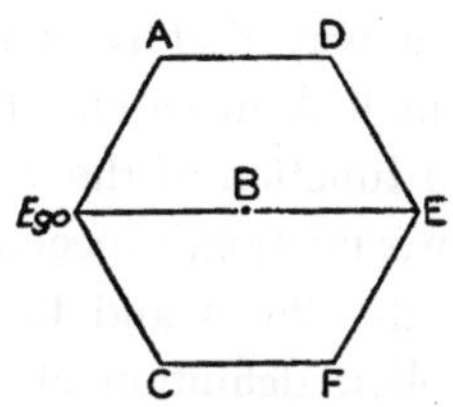

FIG. 104

which gives quite another picture, with the clear centering about *B*.

Fourth Description

With the qualifications as to direction the schema becomes

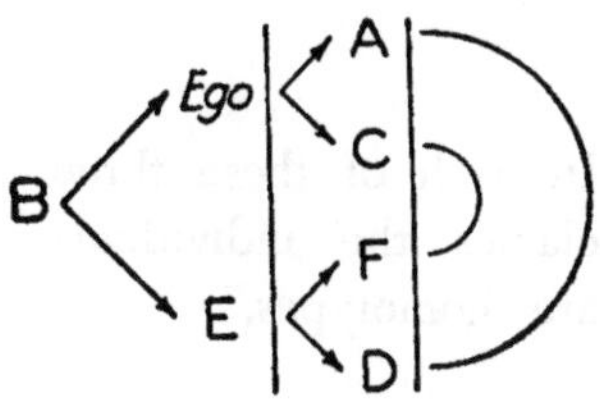

FIG. 105a

or

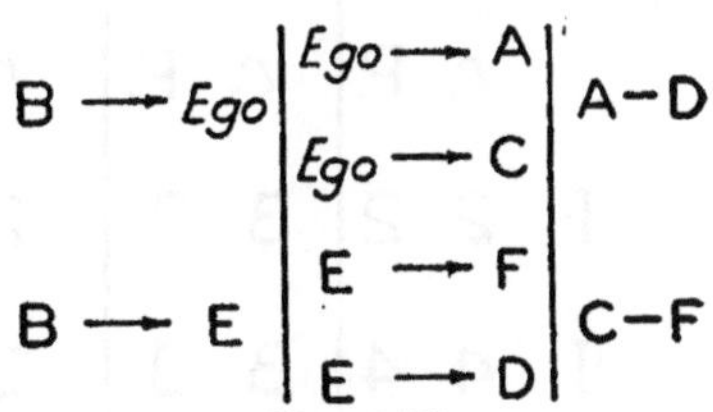

FIG. 105b

and the diagram

FIG. 106

Everything is now structurally clear; there is no confusion as in the previous descriptions, schemas, and diagrams. The girl *had* stated all the relations involved, but in a rather disturbing, miscentered aggregate. The first and the second descriptions and the corresponding figures go against the objective structure of the situation: they are blind to it, they violate it, they center it wrongly. There is nothing to be said against *starting* with the Ego in the description, but in a reasonable description one would not center all that follows about it. On the contrary, one would see and report the Ego in its (secondary) place within the structure.

If the reader compares the first and the second with the third or the fourth diagram, he will see what I mean when I say that in the first and the second the succession, the grouping, the centering are done in a way that is blind to the structurally adequate picture. But let us look more closely: if we wish to characterize the place, role, and function of the various parts of the picture, we may, in a first approximation, proceed in the way in which the logicians characterize the items and the relation-lines in a relational network. The implicit definition of points by their relational places in the network, in terms of the number of relations (r) to immediately neighboring points (P_{I}), to secondary points (P_{II}), etc., in our example gives the following schema:

	B		Ego, E		A,D,C,F	
	r.	P.	r.	P.	r.	P.
I	2	2	3	3	2	2
II	4	4	3	3	3	3
III	2	–	2	–	1	–

In each of these three classes the individuals are "homotypes."

Fig. 107

In the third description, and at the level I, we have:

for *B* for *Ego* for *D*

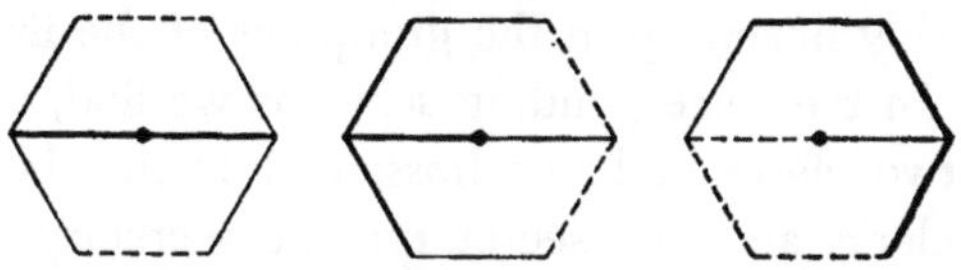

Fig. 108

while the first description gives:

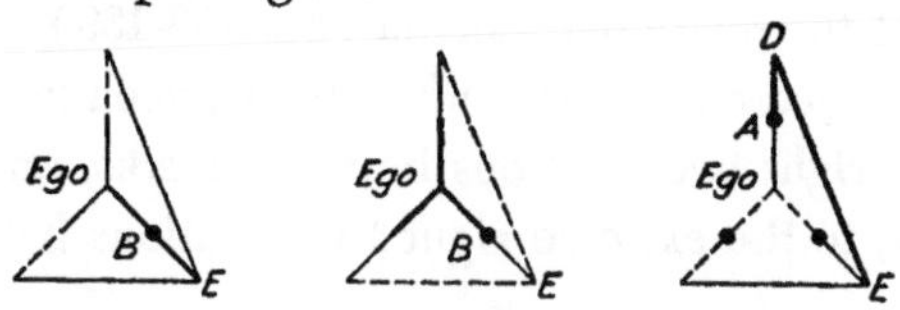

Fig. 109

If we characterize individuals of the three classes by the letters α β γ we get:

in the first description

1	2	3
	γ	γ
β	α	β
	γ	γ

Fig. 110

and in the third description

1	2	3
	β	γ
		γ
α		
	β	γ
		γ

Fig. 111

With regard to the relations, we get:

in the first description

1	2	3
$\beta\gamma$	$\gamma\ \gamma$	$\gamma\ \beta$
$\beta\alpha$	$\alpha\ \beta$	
$\beta\gamma$	$\gamma\ \gamma$	$\gamma\ \beta$

Fig. 112

and in the third description

1	2	3
$\alpha\beta$	$\beta\ \gamma$	$\gamma\gamma$
	$\beta\ \gamma$	
$\alpha\beta$	$\beta\ \gamma$	$\gamma\gamma$
	$\beta\ \gamma$	

Fig. 113

The result is similar if we take into account the directed character of the relations in the second and fourth descriptions.

We find that in this example there are three kinds of people; we may call them boss, secretaries, clerks. Centered about the Ego, they are considerably mixed up in the groupings of the first and second descriptions. In the corresponding schemas we find, as members of one group, two clerks and one boss; and as members of another group, two clerks and one secretary; one secretary, the girl, is a single item on the left side, whereas the other secretary is one member of a triplet on the right side. Similarly with the relations: On the left side of the relational schema (pp. 138-139) we have two relations between secretary and clerks, one between secretary and boss; further to the right two relations between clerks, one between boss and secretary; to the extreme right two relations between secretary and clerks.

In contrast, everything is in clear, surveyable, flawless order when we turn to the third and fourth descriptions and schemas (p. 140). The starting point is the boss, then come the two secretaries, and finally the four clerks. The first two relations are the relations between boss and secretaries, then follow the relations between secretaries and clerks, and finally the relations between the clerks.

In short: Viewed piecemeal, the first and second descriptions are in a way correct, and complete in all details, but they introduce a centering, a grouping that is blind to, and in violation of, the logical hierarchy in the situation. They put together things which are different in nature and let things appear as structurally different which are not. Seen from the girl's subjective point of view, i.e., without taking her secondary position into account, the description distorts the structure; it fails to do justice to the structural meanings of the parts.

What we have just said concerning the difference between miscentered description and structurally appropriate description applies even if the situation is characterized in terms of implicit definitions, relation-numbers, and so forth. (Nevertheless, the question of clearness of classes as expressed in such terms is not the real heart of the matter. The reader will be aware of the fact that we must exercise caution when we use this logistical method.)

Suppose the owner of the business were to come into the picture.

The relational network would be:

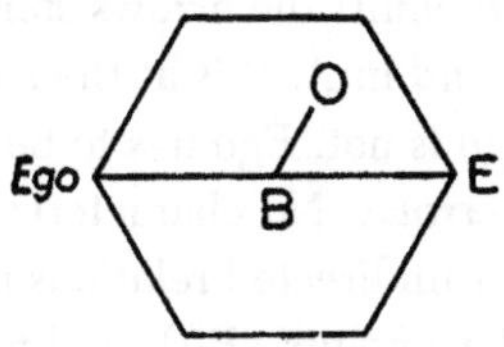

FIG. 114

In terms of relational numbers, B may then easily appear as homotype with Ego and with E, since all three have the same implicit characterization.

	B	Ego, E
	r P	r P
I	3, 3	3, 3
II	4, 4	4, 4
III	2, –	2, –

FIG. 115

These numbers do not tell the whole story. For instance, under II there is a difference in the meaning which the number 4 has in the various cases: for *B* it means 2 + 2, for *Ego* and *E*, 1 + 2 + 1.

In cases such as this, it is decisive how the directions of the relations are distributed. What of a situation in which there were a boss, two secretaries, and one clerk?

FIG. 116

Without taking into account the directedness of the relations, Ego and B would be real homotypes:

FIG. 117

Here we see the decisive point: the arrows *start* from *B*, *go* on the left side *through Ego* and *end* in *A*. *B* is at the source of the arrows, the center in this sense; *Ego* is not. *Ego* has to be viewed in its place within the pattern of the *arrows*. No characterization suffices which is in terms of the number of undirected relations to immediate neighbor points (P_{I}), to secondary points (P_{II}), and to all points in terms of numbers, of relations, etc.

The deeper meaning of being the center does not rest on the fact that what is single is outstanding; it is more important that the center is the source of the arrows, that it is the heart of the matter. There could be, for instance, two bosses, partners, and one secretary, and still this secretary, Ego, would not be the center in terms of the business structure. (It could be, of course, in other regards, say, if the behavior of both bosses were determined by the wish to marry her.)

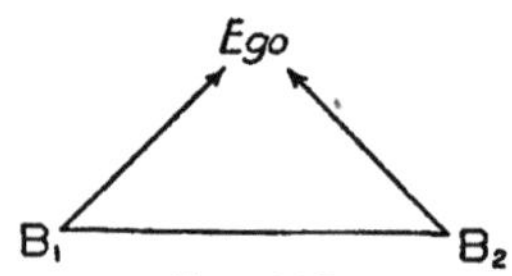

FIG. 118

Centering, then, is not a matter merely of the distribution of numbers of relations; it is a matter of the inner structure of the whole picture, which emerges only if the direction of the relations within the whole pattern is given, and thus the functional meaning of each term.

How do people react to inadequate descriptions? The girl's first description led to the impression that she was the center. "Are you the boss?" I had asked. This question assumed that the arrows were in line with her description.

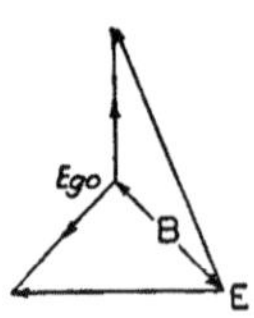

FIG. 119

The second description was at first puzzling because the distribution of arrows did not fit the conceived structure (p. 139). In terms of the arrow distribution *Ego* no longer appears unperturbed as center. At the same time *B* becomes conspicuous as being at the source of the vectors.

If a subject looks at this diagram. a process of the following kind often develops: "This looks funny. How complicated in the center (*Ego*), with those arrows! What funny places for the lines without arrows!" Then some subjects suddenly center on *B*, saying: "The picture is distorted. Referred to the line which passes through *B*, the diagram looks like two wings which need some adjustment . . .", and a new diagram is drawn. Others discover first that "the funny complicated constellation of arrows in *Ego*—'two out, one in'—is structurally repeated in *E*," and in this way reach the view of our last diagram.

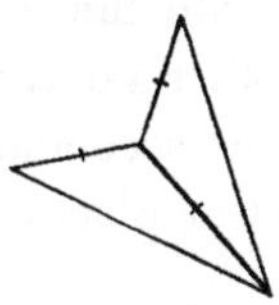

FIG. 120

The preceding diagram clearly appears as something funny, something that is not in order, something that must be improved. Then a strong dynamic process follows which changes that diagram into the structurally clear view of the last diagram.

Some subjects—fewer in number—already show similar reactions when they are confronted with the diagram of the first description,

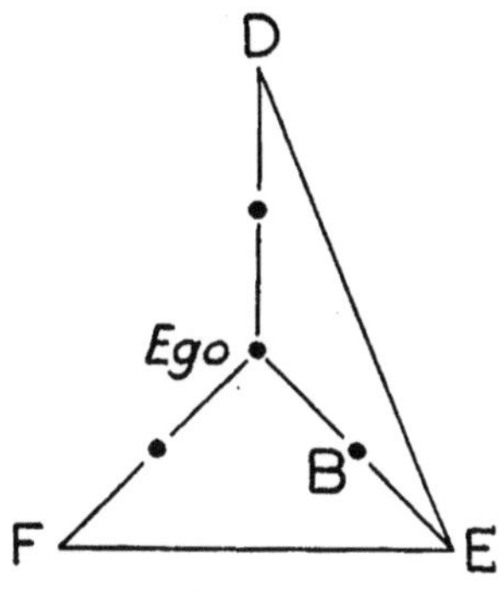

FIG. 121

and when they receive a negative answer to their question whether there is not a line FD. On remembering that the length of lines in relational networks is indeterminate, they sometimes get the "two-wing" view and thus reorganize the picture as in Fig. 119.

In experiments without diagrams, when only the girl's second description or the symbolic list of directed relations is used, reactions are not so strong, not so clear, but sometimes positive nevertheless. Looking for the person who gives orders and does not receive orders, the subjects focus on *B*, and thus recognize the girl's intermediate position.

Often in these cases, too, the transition appears to be a process of restructuring, although this is not so clear as it is with the diagrams: at first *Ego* is conceived as center with most other items in peripheral shadow; then *B* is focused, and *Ego* gets its secondary place, while the other items still do not occupy very clear places in the structure.

Especially in the last mentioned cases the "new idea" has the appearance of a "hunch" because of the initial lack of clearness to which too many details greatly contribute. The reasons for the guess are often not explicit but have merely the form of a direction of centering. In the experiments with the diagrams, too, there is at first an unclear state of affairs, then "some vague idea" emerges which involves directions of recentering, until suddenly the picture crystallizes into the complete new structure.

CHAPTER V

Finding the Sum of the Angles of a Polygon

1. During a conversation at lunch about ornaments, the discussion touched on closed geometric figures such as triangles, rectangles, hexagons, and other polygons. At one point a friend, an artist, remarked: "The sum of the angles of all such figures must of course be the same in every case." Everyone laughed. I found myself in a remarkable situation. I said, "Of course the sum of the angles is *not* the same. For the triangle it is 180°, for the rectangle, 360°, for the hexagon, 720°. But I feel the assertion must be right somehow, it touches an essential point." This feeling continued strong in me. On the one hand, it was clear that the sum of the angles was not the same in the various polygons; on the other hand, I had the feeling that I could not let the matter rest; it had to be cleared up, there had to be a way of seeing through it. There was a beautiful point in it but I did not know how to get at it. It was impossible to understand or even to express just where the problem lay. The compulsion remained: "There must be a solution. What on earth is the matter?"

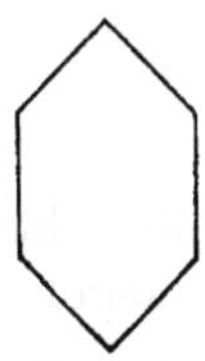

FIG. 122

For the others taking part in the conversation, there was no trouble. The matter was settled when they saw that the assertion was obviously false.

In the hours that followed, in which I had other things to do, the problem continued to be active. After several hours it had developed to this stage: "There is, on the one hand, the sum of the angles (A) in a figure; on the other hand, there is the closed completeness (B) of the figure. Between A and B there is only an "and," a simple conjunction. There is the one, here is the other. This *and*-relationship between them is somehow blind. What is really behind it all? What is the trouble? The two must have something to do with each other. It was not a feeling of two contradictory propositions. I was directed toward the positive problem, "How can I comprehend?"

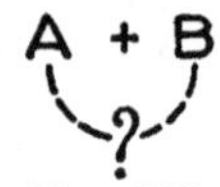

FIG. 123

2. The following day, while I was engaged in other work, this idea suddenly came to me, vague, indefinite and uncertain, something

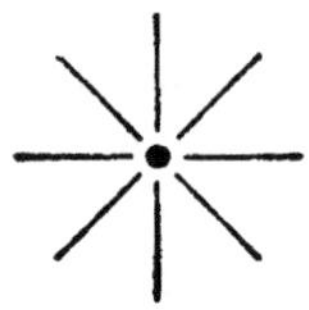

FIG. 124

like this: Here is a *point*. Around a *point* there is a complete "angular space" of 360° (one whole angle). Must there not be

FIG. 125

something similar to this in the case of the closed figure? But this most hazy notion could not be further clarified at the time.

Three days passed. Whatever I did, there was always present this same strong feeling, the feeling of something unfinished, of being directed toward something I could not grasp. I felt repeatedly that I could almost tell where the trouble lay, upon what it depended, in what direction the solution would develop, but it was all in a colloidal, indefinite state so that I could not formulate it concretely. Many times it seemed so clear that "I need only write it down," but when I attempted it, I could not, the ideas could not really be formulated.

(I have found a similar course of development in many truly great intellectual accomplishments: this same feeling of directed tension with the real situation hazy, colloidal. Somehow the form the solution will take, is in its essentials on the tip of one's tongue, but it cannot be concretely grasped. It is a condition which may often last for months with many days of depression during which little success is evident; yet one cannot drop the matter.)

3. Again after two days the following question arose: "If I have a point, there is a complete angle around it. If I have a *straight line*, there is also an angular space around it. Now, if I have such a

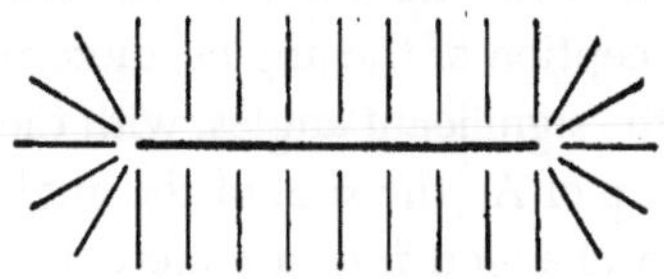

FIG. 126

straight line, *how must I proceed to obtain a closed figure*? Simply by continuing the straight line? Not at all. I must *break* the line at some point, if it is ever to form a closed figure." This led quickly to

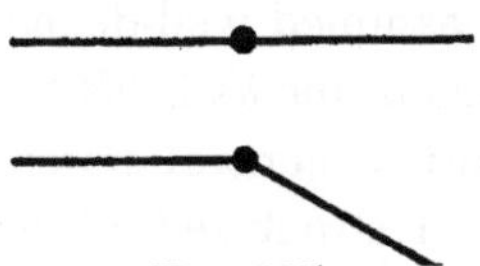

FIG. 127

the idea: "Let me look first at the sum of the exterior angles." And what happened? In breaking, the straight angle broke into two "side angles," each of them a right angle and between them arose a delta (δ), an "angle of rotation." It is the δ's, the rotation,

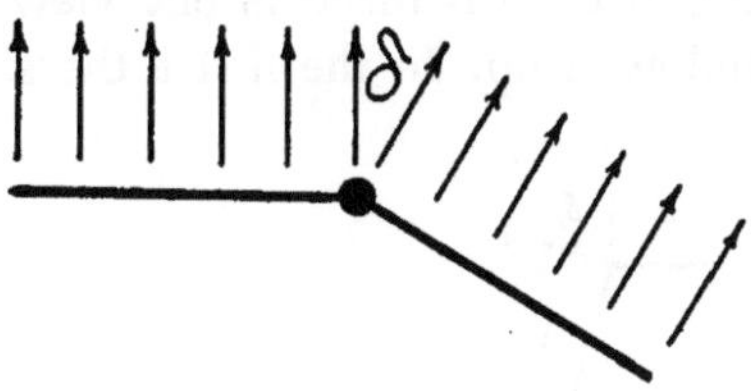

FIG. 128

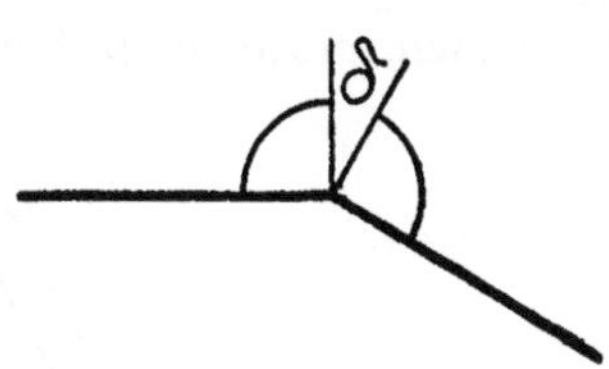

FIG. 129

that are the important thing. And in the whole figure, as it closes, the sum of the δ's must be . . . a complete revolution, an angle of 360°, no matter how many side angles the figure has!

Every side has two exterior right angles, one at each end. There can be as many sides, therefore as many such angles, as one wishes; but in every figure the δ's, the angles of rotation, must make a complete revolution. This was an "intuition." I was very happy at this moment. I had the feeling: "Now I see through the matter."

What actually happened? I started with the usual notion of the angles, and the completion or closure. In trying to see how the closure comes about, the total exterior angle of a corner became two right angles plus δ; the two right angles became peripheral to the issue; the δ was envisaged with the other δ's as forming a complete rotation. With this conception of the angles, there was suddenly an interlocking of the δ's, the significant angles, with the closure of the figure. The and-relationship of A (the sum of the angles) and B (the closed completeness) had changed into a coherent, comprehensible, lucid togetherness. A and B were no longer just pieces set up alongside one another, they were now parts of an inner togetherness. Closing the figure required the δ's to complete 360°. This process of integration served as the solution: what had been just an obscure and unsatisfactory sum, now assumed well-defined shape.

The idea that the sum of the δ's is 360° came into existence not as some assumed proposition, general statement, or belief, but as an "intuition": seeing in a structured view of the figure the inner relatedness between closedness and all the δ's.

Quickly this crystallized into ways of proceeding:

1) There was the realization of what happens—what reasonably must happen—if I go around the figure step by step from the first side of a first δ: to close the figure, I must reach the starting line again, complete a full turn. This had been realized in one view at first;[1] then, proceeding, it was followed up: In the first δ the side

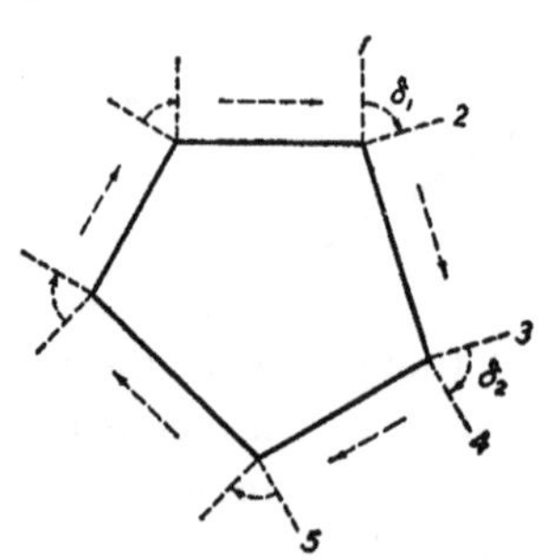

FIG. 130

[1] Long afterward I found in a book a remark, made by the physicist Ernst

of the angle turns by a certain amount into the position of the other side, 2, moves to 3—with *no* turn, as 2 and 3 are parallel; turns from 3 to 4—and so on. In going around, in closure, in reaching position 1 again, the side must have made altogether the full turn of 360°.

2) Just after that came: let the side lines of the figure shrink in size, to zero length. What happens? The distance between neighboring parallel sides of the side angles decreases, these lines fall into identity, so do the corner points—and I have just the picture shown below: the point with the angular space of 360° around it, built up out of the δ's!

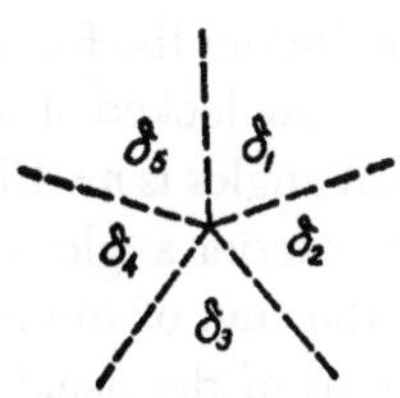

FIG. 133

3) Here the following question came up: what about figures in which a corner is indented, so that the angle does not show the clear

---

Mach, which showed a somewhat similar step in breaking the line. In this way,

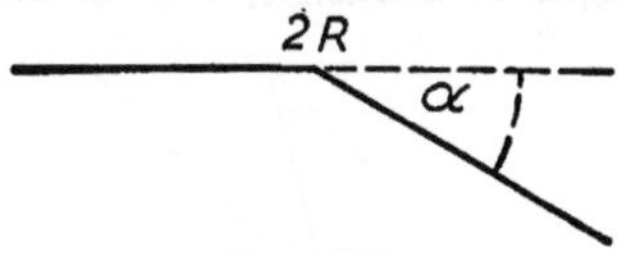

FIG. 131

too, the complete 360° of the sum of the δ's is reached. The view is somewhat

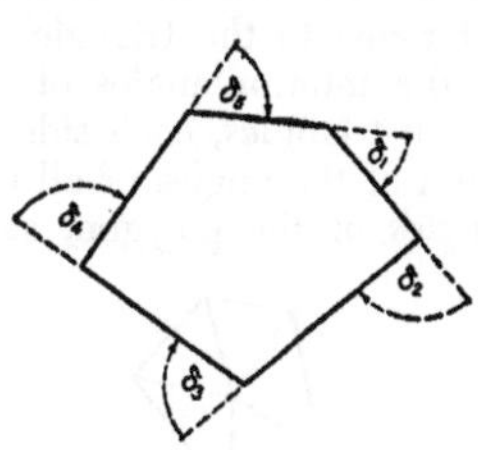

FIG. 132

different: here, the angle is not divided into R, δ, R but into $2R$, δ which makes the conditions for envisaging the completeness of the turn somewhat different psychologically.

structure of the side angles with the δ between? In putting the ques-

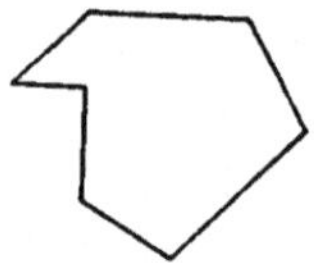

FIG. 134

tion the answer was clear: it does not matter at all; if the side of the angle turns back it has to make up for this and altogether the δ's must complete the 360°.

4) The usual method of finding the formula for the sum of the external angles of a polygon now looked strange indeed: "The sum of all the internal *and* external angles is $n \cdot 4R \ldots . \Sigma\, i + \Sigma\, e = n \cdot 4R$." Therefore, the sum of the exterior angles is $n \cdot 4R$ minus the sum of the interior angles. Since the sum of the interior angles is known to be $n \cdot 2R - 4R$ on the basis of the usual proof with the triangles,[2] we reach the formula $\Sigma\, e = n \cdot 4R - (n \cdot 2R - 4R)$. Carrying out the subtraction, we get: $n \cdot 4R - n \cdot 2R + 4R = n \cdot 2R + 4R$.

In this formula the $n \cdot 2R$ is the outcome of subtracting the $n \cdot 2R$ of the triangles from the $n \cdot 4R$; the $4R$ is the outcome of a second

[2] The sum of the angles of a triangle, 180° or $2R$ (two right angles), is

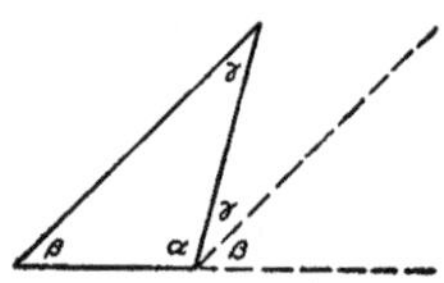

FIG. 135

usually derived with no reference to the triangle's being a closed figure. The usual proof of the sum of the interior angles of a polygon is the following: Inside the polygon construct n triangles, each side of the polygon forming the base of one triangle. The sum of the angles of all the triangles is $n \cdot 2R$. To get the sum of the internal angles of the polygon, subtract from $n \cdot 2R$ the con-

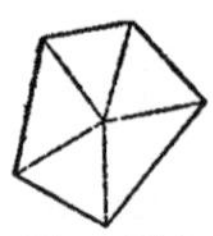

FIG. 136

tiguous angles of the triangles which lie around the middle point. The latter are $4R$. Therefore $\Sigma\, i = n \cdot 2R - 4R$.

negation of the negative $4R$ in the formula for the internal angles. The terms are understood in a roundabout way with no direct relation to how the figure is closed by the angles of the polygon.[3] But meanwhile I have seen what $n \cdot 2R + 4R$ really represents, viz., the sum of the side angles, i.e., the two right angles attached to each side, $n \cdot 2R$, plus the complete revolution, $4R$, the closure, accomplished by the δ's.

5) At this moment there came a curious thought: Why do we call a triangle just a triangle? Why don't we call it, for example, a quad-

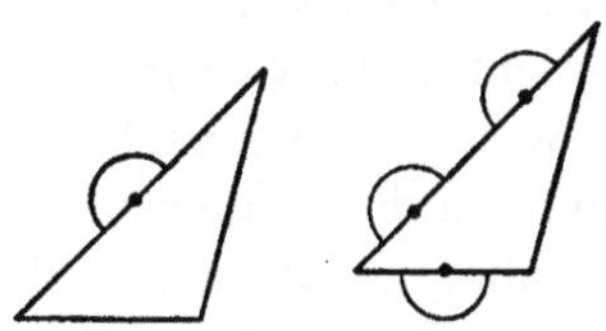

FIG. 137

rangle or a hexagon? We certainly could, because in point of fact there is an angle at every point in the sides. These are not counted. Why? Is this an arbitrary matter? No. Now it is clear: at these points in the sides there are no δ's. These places contribute nothing to the breaking of the line that bounds the figure and to the return to its beginning, to the closing of the polygon by the rotation of the δ's.

6) But how about the internal angles? When I faced this question now, I had again no notion how it could be answered directly. Again a vague idea developed first: there is a complete angle of 360° around a point; also around a figure. On the *inside*, the figure is . . . a "hole"! Quickly this became clear: there must be a complete *negative* angle of 360°: internally *the side angles overlap*. The

[3] It is true that the $4R$ in the formula for the internal angles is directly related to closure in the sense of the fitting together of the triangle tops; but the inner relatedness between the sum of the angles of the triangle itself and its closedness is not equally clear.

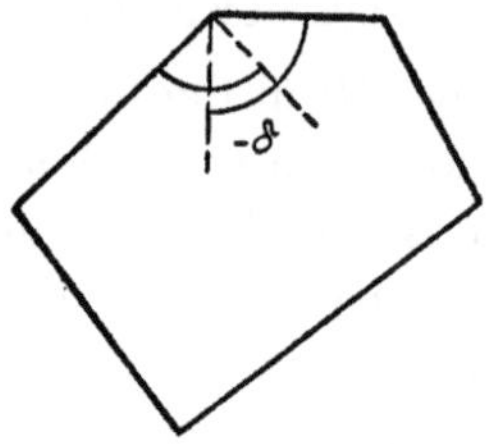

FIG. 138

amount of this overlap represents a negative angle of rotation, $-\delta$. The sum of these angles, when the figure closes, must be equal to a complete negative angle of 360°.

Here the reader might well ask what came out of all this. Exactly the same formula as was known before, but in a new light: the terms of the formula acquired a direct functional meaning.

And this understanding led at once to the joyful insight: if the side angles and the particular number of them are peripheral, if only the rotation of the δ's is relevant, then the same applies to any closed *curve* in a plane, to the closure of a circle, an ellipse, etc.—(I omit the derivation).

7) But the problem was still unfinished. As the solution became clear, a requirement became urgent: if this line of thought really goes to the heart of the matter, then it should also hold for *any* closed figure. It should hold for three-dimensional solids, for four-dimensional, *n*-dimensional, indeed for all closed figures . . . with the necessary changes for non-Euclidean space.

In six weeks of hard work I succeeded in actually understanding three-dimensional figures. (A year later I heard that a mathematician had found the formula for solids years ago, yet I would not have liked to miss these experiences which led to real insight.) During these weeks the problem was constantly active, it continued to work. I studied concrete solids like the cube, parts of the cube, certain pyramids, etc.; the nature of the solid angles with regard to the way, together, they are parts of a complete solid angle. During this time I developed considerable ability to visualize solid angles and to fit them together in imagination. There was no trial and error with regard to formulas, no trying of hypotheses; there was rather the realization of what happens if the solid angles of an imagined, concrete solid are put together at one point; for instance, how the angles of a cube, viewed together in the center of a sphere, make one complete solid angle;[4] what sums the angles of other solids give —parts of the cube, pyramids, parallelepipedons, etc.

There were very dramatic moments as, for instance, when a friend said, "Stop worrying so much about this. The problem is insoluble, for the sum of the angles of a pyramid varies if you alter the altitude. To be sure, it varies continuously."

8) But the process went right on. After a great deal of effort, the solution for three-dimensional solids came in a half-waking moment during the night. Although I could not recall having written anything, in the morning I found the following formula on a slip of paper:

$\Sigma e = \Sigma$ plane angles $+ \Sigma$ edge angles $+ \Sigma \delta$ $(= 1)$, where e stands for an external solid angle. Take a plane (a); break it along a straight line (b); erect normal planes to each plane (c). Between

[4] Just as, in two dimensions, a square angle gives one-fourth of the whole angular space, all four completing it, or the angle of a regular hexagon one-

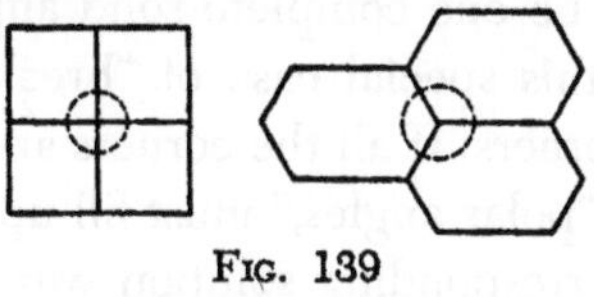

Fig. 139

third, three completing it. In general, teaching what an angle is should bring the view of the angle as part of the whole angle, or of the complete rotation together with the remainder (see Chap. III, p. 109).

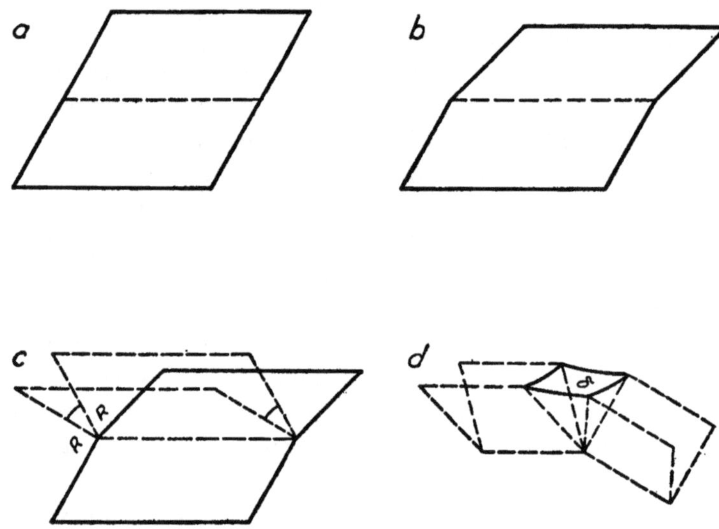

FIG. 140

the normal "plane angles" (corresponding to the side angles, *R*, of the two-dimensional figures), you have an "edge angle" (c); break this at one point (d), and you get the δ. For closure of the solid the sum of the δ's must be one complete solid angle! It was soon seen that what holds in this special case of "breaking the plane" holds generally for solid corners. If all the corners are seen as the center of a sphere, the δ's the "polar angles," must fill up the sphere. This was followed up. The corresponding solution was then reached for the sum of the inner angles, involving the solid "hole," the −1 of the sum of the δ's.

The following days were devoted to exact demonstrations, including spheres, etc.

I shall not follow up the line of my further thinking. I shall stop here in the report, with this joyful moment in which the inner relation between closure and the sum of angles became translucid in solids as in plane figures.

Summarizing:

1. The feeling of a significant interrelatedness between the structure of closed figures and the sum of their angles, and the need to grasp it in a clear fashion.

2. A primitive whole-idea of closedness and "angular space." Here was a changing of the goal: instead of looking for the internal angles, the quest for the sum of the external angles was faced, in a hazy feeling that this was structurally the simpler question. (Later this feeling found its clear justification in the process.)

3. Emphasis on a necessary step in the procedure of closing a figure brought a radical change in the meaning of an angle, the creation of the "rotating angle," the δ, through separation of what is structurally relevant to achieving closure from what is not.

4. With the view of the δ's as a real whole, the intuition was reached of the inner relation between angles and closedness. In contrast to the and-sum of the usual angles, the whole of the δ's gives the complete shape, the closure, in the uniqueness of the 360°. Here regrouping is involved.

Fig. 141

The δ parts, separated from the side angles, were viewed as a whole. But even if one were shown the angles with the dividing lines already drawn, so that each angle consisted of three parts, one could remain structure-blind, group in the usual way, (the three angles in each usual angle belonging together, and the sum of the angles still consisting of the usual angles). Here the actual grouping (the separation of the δ's from the structurally peripheral side angles which do not contribute anything to the closing) was directed by the requirement to understand the closedness of the figure. Focusing on the δ's and viewing them as a whole emphasized the structural transposability of this factor (see p. 152) with regard to the structurally peripheral number of side angles, of usual angles, of lines and corners.

5. In order to secure the envisaged result, the matter was followed up in its details, and detailed proof was given. By decreasing the sides to zero, the direct connection of the external angles with the first view of the "angular space" around a point was concretely realized.

6. A difficulty arose and was cleared up; the principle was found

to apply also to the special case in which a corner of the polygon is indented (cf. p. 153).

7. The insight reached, now gave the actual reason for a customary procedure, which in itself falls short of understanding. The customary formula appeared in a new and deeper sense: the terms found a direct, clear, functional meaning.

8. Now the question of the internal angles was faced. Again a global whole-idea arose first: the idea of the complete "hole," the 360° of the sum of the negative δ's.

9. The applicability of this view broadened; it was found to encompass any closed curve in a plane. In consequence of the insight reached, a limitation in the usual view fell away.

10. The need was felt to follow the matter up: if the insight was fundamental, the relation should also be true in three-dimensional figures, etc. Orientation was first sought in solid angle sums. The material was studied in comparatively simple cases of solids. In spite of complexities, the angles were fitted together and their sums found in imagination. At first no possibility was seen of a radical, general solution.

11. The solution appeared one night—structurally clear, as in the very much simpler case of the two-dimensional figures.

The essential thing in the process was clearly the drive to get at the inner structure of the task. Again we see the working of the whole-view, reorganization, regrouping, getting at the functional meaning of parts in the whole, etc., all in the light of the structural requirements.

Each step was a step in a consistent line of thinking; there were no arbitrary steps, no blind trial and error.

That there were several steps in the procedure, that the process did not proceed easily, seems clearly to have been caused by the fact that the process had to overcome customary, in themselves clear, strong, structural factors; and that later, in three dimensions, mastery of a structurally highly complex situation had to be achieved.

CHAPTER VI

A Discovery by Galileo

How did Galileo make the discovery that led to the law of inertia and thus to the beginning of modern physics?

There have been numerous discussions as to how Galileo actually proceeded. The question is not entirely cleared up even now. Historically, in its details, it is a complex question. The situation in which Galileo found himself was burdened with highly complex concepts and speculations as to the nature of movements.[1] Historical interpretations differ on several points, including the extent to which the old concepts played a role in Galileo's thinking process.[2]

Discussions centered on such points as: Was Galileo's thinking governed by induction? Or by abstraction? By empirical observation and experiment or by a priori presuppositions? Was it his principal merit to have made qualitative observations quantitative?

If one studies the literature—the ancient treatises on physics and those of Galileo's time—one finds that one of the great features in Galileo's thinking was his ability to get such a clear, clean structural

[1] "Natural" and coerced movements were distinguished in a specific way. There was the concept of the—necessarily declining— "vis impressa" (imposed force), and speculations about the role of the medium in delaying a body's coming to rest. There were the ideas about the "natural" circular movements at constant speed, etc.

[2] Readers who are interested in the historical development may read: von E. Wohlwill, "Die Entdeckung des Beharrungsgesetzes," *Zeitschrift für Völkerpsychologie und Sprachwissenschaft*, 1883, Vol. XIV, pp. 365-410; 1884, Vol. XV, pp. 70-135; Ernst Mach, *Die Mechanik in ihrer Entwicklung* (F. A. Brockhaus, Leipzig, 1908); Alexander Koyré's beautiful studies, *Etudes Galiléennes* (I, II, III) (Hermann, Paris, 1939); and, of course, first of all, the writings of Galileo himself.

insight against the highly complex and sophisticated background.

I shall not attempt here a historical reconstruction. This would require a thorough discussion of much source material—and I am no historian. Besides, the printed historical material would not be enough for the psychologist, who is interested in characteristics of the growing process of thought which is not usually put down in writing. Unfortunately we cannot question Galileo himself about the actual development of the process. I should have loved to ask him about certain points in particular.

I shall try to tell the story briefly, in a way that may show some factors which seem to me essential, some lines of the beautiful process. The story that follows is, in certain respects, merely a psychological hypothesis with no claim to historical correctness, but one from which, I think, something may be learned for our problems.

I propose that the reader not only read what I am going to tell, but that he also try to think with me.

I

Here is the situation:

1. If you hold a stone in your hand and let it go, it will fall down. Heavy bodies do. The old physics said: "Heavy bodies tend toward their home, the earth."

2. If I give a push to a body, say a carriage, or if I roll a ball straight ahead on a horizontal plane, it will move, will continue moving for a while, will then come to rest—sooner if I push it gently, somewhat later if I push it hard.

This is the simplest meaning of the old *vis impressa.* "The moving body sooner or later comes to a standstill if the force which is pushing it no longer acts."

Isn't that true? It is obvious.

3. And there are, of course, several additional factors to be considered in connection with questions of movement: e.g., the size of the object, its form, the surface along which the body is moved, the presence or absence of obstacles, etc.

So we know a great many facts about movement. They are familiar to us. Do we understand them? It appears so. Do we really know how movement comes about? Do we see the principles at work?

Galileo was not satisfied with this knowledge. He asked himself: "Do we know *how* such movements really proceed?" Driven by a desire to get at an understanding of the fundamentals, to get at the inner laws involved, Galileo said to himself: "We know that a heavy body falls, but *how* does it fall? In falling it acquires speed. The speed is greater if the distance it falls is greater. What happens to the speed as the body is falling?"

Common experience gives only a vague picture. Galileo started to make observations and experiments in the hope of finding out what happens to the speed, and whether it is governed by principles we can understand. His experimental setups were very crude compared with those that physicists developed later, but in making these observations and experiments he tried to form and to test an hypothesis. First he made a wrong guess, then he found the formula for the acceleration of a falling body. Since the speed of falling is so great that exact values are not easy to determine, Galileo, desirous of studying the question more thoroughly, had taken counsel with himself: "Could I not study this in a more convenient way? Spheres roll down an inclined surface. I shall study them. Isn't free falling simply a special case, the case of falling at an incline of 90° instead of a lesser angle?"

Studying the acceleration in the various cases he saw that acceleration decreases consistently with the angle of inclination: the order of the angle corresponds with the order of declining acceleration.

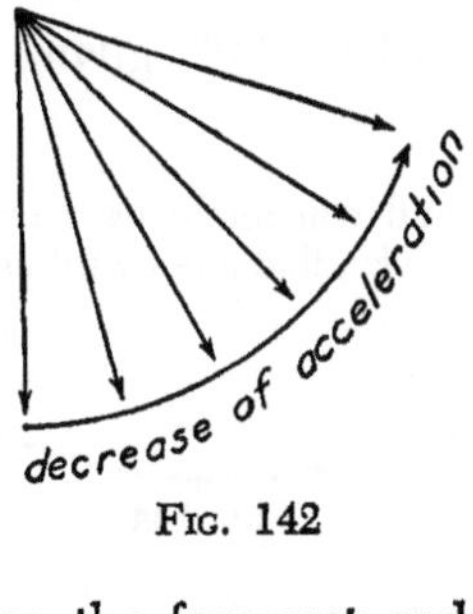

FIG. 142

The acceleration became the foremost and the central fact as he realized the principle connecting the decline of acceleration with the size of the angle.

II

Then suddenly he asked himself: "Is this not just half the picture? Is not what happens when one throws a body upward, when one pushes a sphere in the uphill direction, the symmetrical other part of the picture which repeats, like a reflection in a mirror, what we already have, and which completes the picture?"

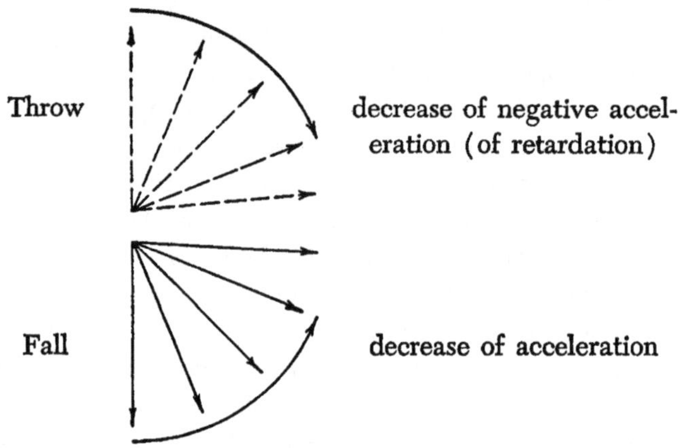

Fig. 143

When a body is thrown up, we have not positive but negative acceleration. The body slows down in the course of its upward movement. But again, symmetrically with the positive acceleration of a falling body, this negative acceleration decreases as the angle of inclination diminishes from 90°, giving a closed, consistent picture.[3]

[3]Galileo had envisaged and concretely used an idea of structural dynamic symmetry in opposite events, viz., that a body rolling down an inclined plane will

Fig. 144

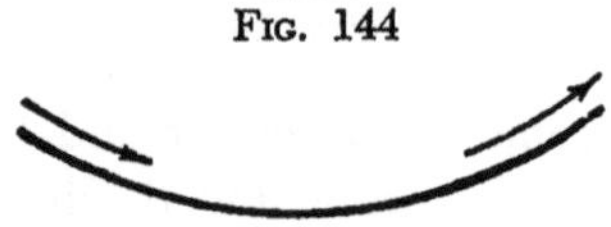

Fig. 145

III

But does this complete the picture? No. There is a gap. What happens when the plane is horizontal, when the angle is zero, and the body is in motion? In all cases we may start with a given velocity. What *must* consistently happen in accordance with the structure?

The positive acceleration below and the negative acceleration above decrease from the vertical cases to . . . no positive or negative acceleration, that is . . . constant movement?! If a body moves horizontally in a given direction, it will continue to move at constant velocity—for all eternity, if no external force changes its state of motion.

This is in direct contradiction to the old statement in point 2 above (p. 161). A body moving at constant velocity will never come to rest if no external hindrances are at work, regardless of whether the forces which brought about the movement were weak or strong.

What an amazing conclusion! Apparently contradicting all familiar experience, yet required by the consistency of the structure.

Of course we cannot perform the experiment. Even if we could eliminate all external hindrances, which we cannot, observation for eternity is out of the question. On the other hand, the decreases in the change of acceleration clearly indicate a zero amount of change for this case.

Galileo's views were confirmed and provided the basis for the development of modern physics.

The modern reader is, of course, acquainted with these views. I shall illustrate with a simple familiar example. The biggest task in moving a train is at the beginning, in accelerating it from rest. Once in motion, provided tracks and wheels are smooth, less power is required of the engine to maintain the speed, the train goes almost by itself. If we now smooth the wheels and the track more and more, and study the decreasing amount of power necessary to run the train, our curves will indicate to our surprise that with the wheels

have to rise to the same height on an opposite inclined plane, its velocity decreasing in the same way that it had increased in rolling down. He had first envisaged such dynamic symmetry in the oscillations of the swinging lamp observed in a church at Pisa.

and track ideally smooth, offering no friction, it would require large counterforces to stop the train, to bring it to rest.[4]

What are the essentials of this process?

First: the desire to clear up, to find out what happens when a body is falling or rolling down; the desire to see whether there is an inner principle in these events; a desire to view them comprehensively at the various angles of inclination.

This centered thought on what happens to acceleration. Experimental setups came into existence through the guess that centering on the question of acceleration might bring structural clarification.

The various cases appear as parts in a well-ordered structure, showing the relation between the angles of inclination and the amount of acceleration. Each of the cases has its place in the group, and what happens in each case is understood as required by this place.

Secondly, this structure is now seen as part of a broader context: there is another complementary part, understood as symmetrical to the first, forming a whole, in which the two halves are the two big corresponding subgroups, with positive acceleration in the one, negative acceleration in the other. The whole-qualities of the two halves complete each other. They are seen in one view, in their structural symmetry, in their consistent whole-structure.

Thirdly, it appears that there is a critical place in the structure—the place of horizontal movement. The structural principle of the whole provides clear requirements for just this place. In view of these requirements it appears as the case in which no acceleration and no retardation occur—constant velocity.

Rest, then, becomes a special case of constant velocity, a case of the absence of positive or negative acceleration. Rest and constant rectilinear movement in the horizontal direction appear as structurally equivalent.

[4] Compare the much more simplified description of Galileo's thought process given by Einstein and Infeld in *The Evolution of Physics* (Simon and Schuster, 1938) pp. 5 ff.

Of course operations of traditional logic are involved, such as induction, inference, formulating theorems, derivation—and observation and ingenious experimentation. (One of the great features in Galileo's thinking was the way he combined strict reasoning, mathematical procedure, with the use of experiment to test the theoretical idea or to answer a theoretical problem.) But all these operations function in their place in the whole process. The process itself is governed by the recentering which originates in the desire to get comprehensive insight. It brings transformation, seeing things as parts of a new, clear structure.

Fundamental changes in concepts were involved in the transition from the old view to the new. The meanings of the concepts about motion were radically changed in their place, role, and function. The inner connections were viewed in an entirely new structure; a new grouping and dividing took place.[5]

For example, previously, rest and certain "natural" circular movements were fundamentally separated from, opposed to, the other movements. Now rest and constant rectilinear movement are seen together, as structurally equivalent, and separate from, opposed to positively or negatively accelerated movements.

Rising and falling of bodies are viewed together as cases of acceleration in the symmetrical division of the whole picture. Free falling and free rising are viewed as special cases of the whole group of movements in any straight direction.

The end of a movement is no longer seen as the necessary outcome of the diminishing, disappearing effects of *vis impressa* (imposed force). It is now seen in a basically different way: the movement is stopped externally by friction.

Friction is no longer just one of several factors to be considered in movement; it now has a role opposite to that of inertia. While previously the assumption had been that rectilinear motion comes

[5] For the sake of brevity I shall use some formulations that were made only later in their full generality, but that were somehow implied or already foreseen in Galileo's views. Galileo himself was extremely cautious in his formulations.

Galileo's formulation concerns movement in the horizontal. He also applied the principle to movements in other directions. He did not generalize it in the form we now know as the law of inertia, but others did so promptly. We are not sure whether he realized the generality of the principle.

to an end with or without friction, due to the natural extinction of *vis impressa*, the new view looks at friction as the fundamental cause of the limitation of motion.

Force appears essentially as something determining acceleration.

Everything gets its new meaning by virtue of its role and function in the new structure.

The new concepts opened up a most marvelous vista for understanding a huge number of phenomena. They made possible an entirely new understanding of the celestial movements. Newton subsequently could view these movements as generated by inertia, on the one hand, tending toward rectilinear movement, and by gravitation, on the other.

Productive processes are often of this nature: in the desire to get at real understanding, requestioning and investigation start. A certain region in the field becomes crucial, is focused; but it does not become isolated. A new, a deeper structural view of the situation develops, involving changes in the functional meaning, the grouping, etc., of the items. Directed by what is required by the structure of a situation for a crucial region, one is led to a reasonable prediction, which—like the other parts of the structure—calls for verification, direct or indirect.

Two directions are involved: getting a whole consistent picture, and seeing what the structure of the whole requires for the parts.

Telling the story this way, I often had delightful experiences, seeing what vivid, genuine interest was aroused, watching dramatic things happen to the class, seeing how, at the critical point, some students exclaimed: "Now I see!" For them it was a change from knowing a number of things to having their eyes really opened to a deeper understanding in a comprehensive view.

CHAPTER VII

Einstein: The Thinking That Led to the Theory of Relativity

WHAT were the decisive steps in the development of Einstein's theory of relativity? Although this is quite a task, I shall try to make them clear to the reader. A number of issues, such as the question of the ether, the relations to the principles of Galilean "relativity," will be excluded from the discussion. The field which Einstein faced in this gigantic process of thought was very large, since it involved most of the fundamental topics of modern physics, difficult issues unfamiliar to those not acquainted with the complexity of modern physics. Although the following sketch will of necessity be condensed, I hope the reader will gain an insight into the nature of the decisive steps.

Those were wonderful days, beginning in 1916, when for hours and hours I was fortunate enough to sit with Einstein, alone in his study, and hear from him the story of the dramatic developments which culminated in the theory of relativity. During those long discussions I questioned Einstein in great detail about the concrete events in his thought. He described them to me, not in generalities, but in a discussion of the genesis of each question.

Einstein's original papers give his results. They do not tell the story of his thinking. In the course of one of his books he did report some steps in the process. I have quoted him in the proper places in this chapter.

The drama developed in a number of acts.

Act I. The Beginning of the Problem

The problem started when Einstein was sixteen years old, a pupil in the Gymnasium (Aarau, Kantonschule). He was not an especially good student, unless he did productive work on his own account. This he did in physics and mathematics, and consequently he knew more about these subjects than his classmates. It was then that the great problem really started to trouble him. He was intensely concerned with it for seven years; from the moment, however, that he came to question the customary concept of time (see Act VII), it took him only five weeks to write his paper on relativity—although at this time he was doing a full day's work at the Patent Office.

The process started in a way that was not very clear, and is therefore difficult to describe—in a certain state of being puzzled. First came such questions as: What if one were to run after a ray of light? What if one were riding on the beam? If one were to run after a ray of light as it travels, would its velocity thereby be decreased? If one were to run fast enough, would it no longer move at all? . . . To young Einstein this seemed strange.

The same light ray, for another man, would have another velocity. What *is* "the velocity of light"? If I have it in relation to something, this value does not hold in relation to something else which is itself in motion. (Puzzling to think that under certain conditions light should go more quickly in one direction than another.) If this is correct, then consequences would also have to be drawn with reference to the earth, which is moving. There would then be a way of finding out by experiments with light whether one is in a moving system! Einstein's interest was captured by this; he tried to find methods by which it would be possible to establish or to measure the movement of the earth—and he learned only later that physicists had already made such experiments. His wish to design such experiments was always accompanied by some doubt that the thing was really so; in any case, he felt that he must try to decide.

He said to himself: "I know what the velocity of a light ray is in relation to a system. What the situation is if another system is taken into account seems to be clear, but the consequences are very puzzling."

Act II. Light Determines a State of Absolute Rest?

Would operations with light lead to conclusions different in this respect from conclusions from mechanical operations?[1] From the point of view of mechanics there seems to be no absolute rest; from the point of view of light there does seem to be. What of the velocity of light? One must relate it to something. Here the trouble starts. Light determines a state of absolute rest? However, one does not know whether or not one is in a moving system. Young Einstein had reached some kind of conviction that one cannot notice whether or not one is in a moving system; it seemed to him deeply founded in nature that there is no "absolute movement." The central point here became the conflict between the view that light velocity seems to presuppose a state of "absolute rest" and the absence of this possibility in the other physical processes.

Back of all this there had to be something that was not yet grasped, not yet understood. Uneasiness about this characterized young Einstein's state of mind at this time.

When I asked him whether, during this period, he had already had some idea of the constancy of light velocity, independent of the movement of the reference system, Einstein answered decidedly: "No, it was just curiosity. That the velocity of light could differ depending upon the movement of the observer was somehow characterized by doubt. Later developments increased that doubt." Light did not seem to answer when one put such questions. Also light, just as mechanical processes, seemed to know nothing of a state of absolute movement or of absolute rest. This was interesting, exciting.

Light was to Einstein something very fundamental. At the time of his studies at the Gymnasium, the ether was no longer being

[1] Cf. below, Act IX.

The layman, not acquainted with modern physics, will not be able to follow the topics under II and III in my brief formulations. While these topics were important in the process, it does not seem absolutely necessary to understand them fully in order to follow the later steps within the positive solution. The lay reader may therefore pass on to Act IV.

thought of as something mechanical, but as "the mere carrier of electrical phenomena."

Act III. Work on the One Alternative

Serious work started. In the Maxwell equations of the electromagnetic field, the velocity of light plays an important role; and it is constant. If the Maxwell equations are valid with regard to one system, they are not valid in another. They would have to be changed. When one tries to do so in such a way that the velocity of light is not assumed to be constant, the matter becomes very complicated. For years Einstein tried to clarify the problem by studying and trying to change the Maxwell equations. He did not succeed in formulating these equations in such a way as to meet the difficulties satisfactorily. He tried hard to see clearly the relation between the velocity of light and the facts of movement in mechanics. But in whatever way he tried to unify the question of mechanical movement with the electromagnetic phenomena, he got into difficulties. One of his questions was: What would happen to the Maxwell equations and to their agreement with the facts if one were to assume that the velocity of light depends on the motion of the source of the light?

The conviction grew that in these respects the situation with regard to light could not be different from the situation with regard to mechanical processes (no absolute movement, no absolute rest). What took him so much time was this: he could not doubt that the velocity of light is constant and at the same time get a satisfactory theory of electromagnetic phenomena.

Act IV. Michelson's Result and Einstein

The famous Michelson experiment confronted physicists with a disconcerting result. If you are running away from a body that is rushing toward you, you will expect it to hit you somewhat later than if you are standing still. If you run toward it, it will hit you earlier. Michelson did just this in measurements of the velocity of light. He compared the time light takes to travel in two pipes if these pipes meet at right angles to each other, and if one lies in the direction of the movement of the earth, while the other is vertical to it. Since the first pipe, in its lengthwise direction, is moving with the movement of the earth, the light traveling in it ought to reach the receding end of this pipe later than the light in the other pipe reaches *its* end.

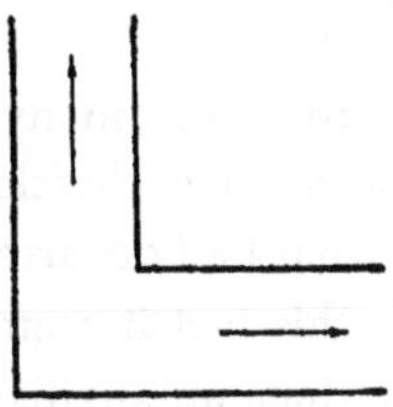

FIG. 146

The arrows above indicate the directions in which the light is traveling. The movement of the earth, and so of the whole apparatus is toward the right.

Actually the arrangement was more intricate. There was at the vertex angle of the two pipes a common mirror, and there were mirrors at the ends. In both pipes light rays from a common source were moving forward and backward, reflected by the mirrors. The difference in time was to be measured by means of an interference effect at the common mirror. (The reader may be inclined to assume that, with the light rays moving forward and backward, the difference introduced by the movement of the earth would be canceled. This is not the case, as some mathematical deliberation will show.) The difference could not escape observation because the measurement by interference was sufficiently delicate to reveal the amount as established by mathematical analysis.

No difference was found. The experiment was repeated, and the negative result was clearly confirmed.

The result of the Michelson experiment in no way fitted the fundamental ideas of the physicists. In fact the result contradicted all their reasonable expectations.

For Einstein, Michelson's result was not a fact for itself. It had its place within his thoughts as they had thus far developed. Therefore, when Einstein read about these crucial experiments made by physicists, and the finest ones made by Michelson, their results were no surprise to him, although very important and decisive. They seemed to confirm rather than to undermine his ideas. But the matter was not yet entirely cleared up. Precisely how does this result come about? The problem was an obsession with Einstein although he saw no way to a positive solution.

Act V. The Lorentz Solution

Not only Einstein was troubled; many physicists were. Lorentz, the famous Dutch physicist, had developed a theory which formulated mathematically what had occurred in the Michelson experiment. In order to explain this fact it seemed necessary to him, as it had to Fitzgerald, to introduce an auxiliary hypothesis: he assumed that the entire apparatus used in the measurement underwent a small contraction in the direction of the earth's motion. According to this theory, the pipe in the direction of the movement of the earth was changed in length, while the other pipe suffered only a change in width and the length remained unaffected. The contraction had to be assumed to be just the amount needed to compensate for the effect of the earth's motion on the traveling of the light. This was an ingenious hypothesis.

There was now a fine, positive formula, determining the Michelson results mathematically, and an auxiliary hypothesis, the contraction. The difficulty was "removed." But for Einstein the situation was no less troublesome than before; he felt the auxiliary hypothesis to be a hypothesis *ad hoc*, which did not go to the heart of the matter.

Act. VI. Re-examination of the Theoretical Situation

Einstein said to himself: "Except for that result, the whole situation in the Michelson experiment seems absolutely clear; all the factors involved and their interplay seem clear. But *are* they really clear? Do I really understand the structure of the whole situation, especially in relation to the crucial result?" During this time he was often depressed, sometimes in despair, but driven by the strongest vectors.

In his passionate desire to understand or, better, to see whether the situation was really clear to him, he faced the essentials in the Michelson situation again and again, especially the central point:

the measurement of the speed of light under conditions of movement of the whole set in the crucial direction.

This simply would not become clear. He felt a gap somewhere without being able to clarify it, or even to formulate it. He felt that the trouble went deeper than the contradiction between Michelson's actual and the expected result.

He felt that a certain region in the structure of the whole situation was in reality not as clear to him as it should be, although it had hitherto been accepted without question by everyone, including himself. His proceeding was somewhat as follows: there is a time measurement while the crucial movement is taking place. "Do I see clearly," he asked himself, "the relation, the inner connection between the two, between the measurement of time and that of movement? Is it clear to me how the measurement of time works in such a situation?" And for him this was not a problem with regard to the Michelson experiment only, but a problem in which more basic principles were at stake.

Act. VII. Positive Steps toward Clarification

It occurred to Einstein that time measurement involves simultaneity. What of simultaneity in such a movement as this? To begin with, what of simultaneity of events in different places?

He said to himself: "If two events occur in one place, I understand clearly what simultaneity means. For example, I see these two balls hit the identical goal at the same time. But . . . am I really clear about what simultaneity means when it refers to events in two different places? What does it mean to say that this event occurred in my room at the same time as another event in some distant place? Surely I can use the concept of simultaneity for different places in the same way as for one and the same place—but can I? Is it as clear to me in the former as it is in the latter case? . . . It is not!"

For what now followed in Einstein's thinking we can fortunately report paragraphs from his own writing.[2] He wrote them in the form of a discussion with the reader. What Einstein here says to the

[2] A. Einstein, *Über die spezielle und die allgemeine Relativitätstheorie* (Braunschweig, 1916) pp. 14 ff. (Editors' note. The passages from Einstein quoted in this chapter were translated by Professor Wertheimer.)

reader is similar to the way his thinking proceeded: "Lightning strikes in two distant places. I assert that both bolts struck simultaneously. If now I ask you, dear reader, whether this assertion makes sense, you will answer, 'Yes, certainly'. But if I urge you to explain to me more clearly the meaning of this assertion, you will find after some deliberation that answering this question is not as simple as it at first appears.

"After a time you will perhaps think of the following answer: 'The meaning of the assertion is in itself clear and needs no further clarification. It would need some figuring out, to be sure, if you were to put me to the task of deciding by observation whether in a concrete case the two effects were actually simultaneous or not.'"

I now insert an illustration which Einstein offered in a discussion. Suppose somebody uses the word "hunchback." If this concept is to have any clear meaning, there must be some way of finding out whether or not a man has a hunched back. If I could conceive of no possibility of reaching such a decision, the word would have no real meaning for me.

"Similarly," Einstein continued, "with the concept of simultaneity. The concept really exists for the physicist only when in a concrete case there is some possibility of deciding whether the concept is or is not applicable. Such a definition of simultaneity is required, therefore, as would provide a method for deciding. As long as this requirement is not fulfilled, I am deluding myself as physicist (to be sure, as non-physicist too!) if I believe that the assertion of simultaneity has a real meaning. (Until you have truly agreed to this, dear reader, do not read any further.)

"After some deliberation you may make the following proposal to prove whether the two shafts of lightning struck simultaneously. Put a set of two mirrors at an angle of 90° to each other ($\vee$), at the exact halfway mark between the two light effects, station yourself in front of them, and observe whether or not the light effects strike the mirrors simultaneously."

Simultaneity in distant places here gets its meaning by being based on clear simultaneity in an identical place.[3]

[3] This involves other problems which we shall here ignore. The reader is referred to Einstein. *Über die spezielle und allgemeine Relativitätstheorie*, pp. 15-16.

All these steps came not by way of isolated clarification of this special question, but as part of the attempt to understand the inner connection that was mentioned above, the problem of the measurement of speed during the crucial movement. In the mirror situation this means simply: what happens if, in the time during which the light rays approach my mirrors, I move with them, away from one source of light and toward the other. Obviously, if the two events appeared simultaneous to a man at rest they would not then appear so to me, who am moving with my mirrors. His statement and mine must differ. We see then that our statements about simultaneity *involve essentially reference to movement of the observer.* If simultaneity in distant places is to have real meaning, I must explicitly take into account the question of movement, and in comparing my judgments with those of another observer, I have to take into account the relative movement between him and me. When dealing with "simultaneity in different places" I must refer to the relative movement of the observer.

I repeat: suppose that I with my mirrors am traveling in a train going in a straight line at a constant velocity. Two shafts of lightning strike in the distance, one near the engine, the other near the rear end of the train; my double mirror being right in the middle between the two. As a passenger I use the train as my frame of reference, I relate these events to the train. Let us assume that just at the critical moment when the lightning strikes, a man is standing beside the tracks, likewise with double mirrors, and that his place at that moment coincides with mine. What would my observations be and what would his be?

"If we say that the bolts of lightning are simultaneous with regard to the tracks, *this now means*: the rays of light coming from two equidistant points meet simultaneously at the mirrors of the man on the track. But if the place of my moving mirrors coincides with his mirrors at the moment the lightning strikes, the rays will not meet exactly simultaneously in my mirrors because of my movement.

"Events which are simultaneous in relation to the track are not simultaneous in relation to the train, and vice versa. Each frame of reference, each system of coordinates therefore has *its special time*;

a statement about a time has real meaning only when the frame of reference is stated, to which the assertion of time refers."[4]

It has always seemed simple and clear that a statement about the "time difference" between two events is a "fact," independent of other factors, such as movement of the system. But, in actual fact, is not the thesis that "the time difference between two events is independent of the movement of the system" an arbitrary assumption? It did not hold, as we saw, for simultaneity in different places, and therefore it cannot hold even for the length of a second. To measure a time interval, we must use a clock or the equivalent of a clock, and look for certain coincidences at the beginning and at the end of the interval. Therefore the trouble with simultaneity is involved. We cannot dogmatically assume that the time which a certain event takes in relation to the train is the same as the time in relation to the track.

This applies also to the measurement of distances in space! If I try to measure exactly the length of a car by marking its end points on the roadbed, I must take care, when I have made my mark at one end, that the car does not move before I come to the other end! Unless I have explicitly given attention to this possibility, my measurements will be misleading.

I must therefore conclude that in every such measurement reference must be made to the movement of the system. For the observer within the moving system will get results which differ from those of an observer in another frame of reference. "Every system has its special time and space values. A time or space judgment has sense only if we know the system with reference to which the judgment was made." We must change the old view: the measurements of time intervals and of distances in space are *not* independent of the conditions of movement of the system in relation to the observer.

The old view had been a time-honored "truth." Einstein, seeing that it was questionable, came to the conclusion that space and time measurements depend on the movement of the system.

Act VIII. Invariants and Transformation

What followed was determined by two vectors which simultaneously tended toward the same question.

[4] A. Einstein, *op. cit.*, pp. 31-32.

1. The system of reference may vary; it can be chosen arbitrarily. But in order to reach physical realities, I have to get rid of such arbitrariness. The basic laws must be independent of arbitrarily chosen co-ordinates. If one wants to get a description of physical events, the basic laws of physics must be invariant with regard to such changes.

Here it becomes clear that one might adequately call Einstein's theory of relativity just the opposite, an absolute theory.

2. Insight into the interdependence of time measurement and movement is certainly not enough in itself. What is now needed is a transformation formula that answers this question: "How does one find the place and time values of an event in relation to one moving system, if one knows the places and times as measured in another? Or better, how does one find the transformation from one system to another when they move in relation to each other?"

What would be the direct way? In order to proceed realistically, I would have to base the transformation on an assumption with regard to some physical realities which could be used as invariants.

The reader may think back to an old historical situation. Physicists in past ages tried to construct a *perpetuum mobile*. After many attempts which did not succeed, the question suddenly arose: how would physics look if nature were basically such as to make a *perpetuum mobile* impossible? This involved an enormous change, which recentered the whole field.

Similarly there arose in Einstein the following question, which was inspired by his early ideas mentioned in Acts II and III: How would physics look if, by nature, measurements of the velocity of light would under all conditions have to lead to the identical value? Here is the needed invariant! (Thesis of the basic constancy of the velocity of light.)

In terms of the desired transformation, this means: "Can a relation between the place and time of events in systems which move linearly to each other be so conceived that the velocity of light becomes a constant?"

Eventually Einstein reached the answer: "Yes!" The answer consisted of concrete and definite transformation formulas for distances in time and space, formulas that differed characteristically from the old Galilean transformation formulas.

3. In the discussions I had with Einstein in 1916 I put this question to him: "How did you come to choose just the velocity of light as a constant? Was this not arbitrary?"

Of course it was clear that one important consideration was the empirical experiments which showed no variation in the velocity of light. "But did you choose this arbitrarily," I asked, "simply to fit in with these experiments and with the Lorentz transformation?" Einstein's first reply was that we are entirely free in choosing axioms. "There is no such difference as you just implied," he said, "between reasonable and arbitrary axioms. The only virtue of axioms is to furnish fundamental propositions from which one can derive conclusions that fit the facts." This is a formulation that plays a great role in present theoretical discussions, and about which most theorists seem to be in agreement. But then Einstein himself smilingly proceeded to give me a very nice example of an unreasonable axiom: "One could of course choose, say, the velocity of sound instead of light. It would be reasonable, however, to select not the velocity of just 'any' process, but of an 'outstanding' process . . ." Questions like the following had occurred to Einstein: Is the speed of light perhaps the fastest possible? Is it perhaps impossible to accelerate any movement beyond the speed of light? As velocity increases, progressively greater forces are required to increase it still further. Perhaps the force required to increase a velocity beyond the velocity of light is infinite?

It was marvelous to hear in Einstein's descriptions how these bold questions and expectations had taken shape in him. It was new, unthought of before, that the velocity of light might be the greatest possible velocity, that an attempt to go beyond that limit would require forces infinitely great.

If these assumptions brought clarity into the system, and if they were proved by experiment, then it would make good sense to take the velocity of light as the basic constant. (Cf. the absolute zero of temperature which is reached when the molecular movements in an ideal gas approach zero.)

4. The derivations which Einstein reached from his transformation

formulas showed mathematical coincidence with the Lorentz transformation. The contraction hypothesis had therefore been in the right direction, only now it was no longer an arbitrary auxiliary hypothesis, but the outcome of improved insight, a logically necessary derivation from the improved view of fundamental physical entities. The contraction was not an absolute event, but a result of the relativity of measurements. It was not determined by a "movement in itself which possesses no real sense for us, but only by a movement with reference to the chosen observation system."

Act IX. On Movement, on Space, a Thought Experiment

The last statement throws new light on the changes in thinking which were already involved in the earlier steps. "By the motion of a body we always mean its change of position in relation to a second body," to a framework, or a system. If there is one body alone, it has no sense to ask or to try to state whether it is moving or not. If there are two, we can state only whether they are approaching or moving away from each other; but, so long as there are only two, it has no sense to ask, or to try to state, whether one is turning around the other; the essential in movement is change of position in relation to another object, a framework, or a system.

But is there not *one* outstanding system in regard to which there is *absolute* movement of a body, "the" space (Newtonian space, the space of the ether), the box in which all movement takes place?

Here I may mention something that happened not just at this point in the development of the process, but may illustrate what was really going on. It transcends the problems of the special theory of relativity: is there no proof of the reality of such an outstanding system? A famous experiment of Newton's had been used as proof: When a sphere of oil rotates it becomes flattened. This is a real, physical, observable fact, apparently caused by an "absolute" movement.

But is this really a demonstration of such an absolute movement? It seems so certainly; but is it actually, if we think it through? In reality we have not a body moving alone in absolute space, but a body that moves within our fixed-star firmament. Is the flattening of that sphere perhaps an outcome of the movement of the sphere relative to the surrounding stars? What would happen if we took a very huge iron wheel, with a small hole at the center, if we suspended in this hole a little sphere of oil, and then rotated the wheel? Perhaps the little sphere would again become flattened. Then the flattening would have nothing to do with the rotation in an absolute space box; rather it would be determined by the systems moving in relation to each other, the big wheel or the firmament on the one hand and the little sphere of oil on the other.

Of course rotation already transcends the region of the so-called special relativity of Einstein. It became basic in the problem of the general theory of relativity.

Act X. Questions for Observation and Experiment

Einstein is at heart a physicist. Thus all these developments aimed at real, concrete, experimental problems. As soon as he reached clarification he concentrated on the point: "Is it possible to find crucial physical questions to be answered in experiments that will decide whether these new theses are 'true'; whether they fit facts better, give better predictions of physical events than the old theses?"

He found a number of such crucial experiments, some of which physicists could and later did carry out.

II

In actual fact, the problem leads on: it led in Einstein's mind to the problems of the general theory of relativity. But let us stop with the story here and ask ourselves: What were the decisive characteristics of this thinking?

The physicist is interested in the relation of Einstein's theory to established facts, in experimental proof, in the consequences for further development, in the mathematical formulas which follow from the theory of relativity in the various parts of physics.

The theorist of knowledge is interested in the ideas of space, time, and matter, in the "relativistic" character of the theory (with all the wrong consequences in the direction of philosophical, sociological, or ethical relativism drawn by others), in the problem of "testability" which played such an important role in Einstein's dealing with simultaneity (and later in the developments of operationalism).

The psychologist, who is concerned with the problems of thinking, wants to realize what went on psychologically.

If we were to describe the process in the way of traditional logic, we would state numerous operations, like making abstractions, stating syllogisms, formulating axioms and general formulas, stating contradictions, deriving consequences by combining axioms, confronting facts with these consequences, and so forth.

Such a procedure is certainly good if one wishes to test each of the steps with regard to its logical correctness. Einstein himself is passionately interested in logical correctness, logical validity.

But what do we get if we follow such a procedure? We get an aggregate, a concatenation of a large number of operations, syllogisms, etc. Is this aggregate an adequate picture of what has happened? What many logicians do, the way they think, is somehow like this: a man facing a work of architecture, a fine building, focuses, in order to understand it, on the single bricks and also on the way in which they are cemented by the mortar. What he has at the end is not the building at all but a survey of the bricks and of their connections.[5]

In order to get at the real picture, we have to ask: How did the operations arise, how did they enter into the situation, what was their function within the actual process? Did they just drop in? Was the process a chain of happy accidents? Was the solution a consequence of trial and error, of mathematical guesswork? Why just these operations? No doubt there were other possibilities at some points. Why was Einstein moving in just this direction? How did it come about that after he made one step, he followed with just that other step?

I shall mention one specific point: How did the new axioms arise? Did Einstein just try any axioms of which certain ones then actually

[5] "I am not sure," Einstein said once in this context, "whether there can be a way of really understanding the miracle of thinking. Certainly you are right in trying to get at a deeper understanding of what really goes on in a thinking process. . . ."

happened to work? Did he formulate some propositions, put them together, and observe what happened until eventually he was fortunate enough to find a proper set? Did such propositions leap into the picture accidentally, and did the changes in the role, place, and function of the items, did their new interrelatedness appear merely as derived consequences?

The technique of axioms is a very useful tool. It is one of the most efficient techniques so far invented in logic and mathematics; a few general propositions provide all that is needed in order to derive the details. One can deal with a gigantic sum of facts, with huge numbers of propositions, by substituting for them a few sentences which in a formal sense are equivalent to all that knowledge. Some great discoveries in modern mathematics became possible only because this extremely simplifying technique was at hand. Einstein, too, used this tool in the accounts which he gave of his theory of relativity.

But, to repeat, the question for the psychologist is: Were these axioms introduced before the structural requirements,[6] the structural changes of the situation were envisaged? Was it not the other way around? Surely, Einstein's thought did not put ready-made axioms or mathematical formulas together. The axioms were not the beginning but the outcome of what was going on. Before they came into the picture as formulated propositions, the situation as to the velocity of light and related topics had for a long time been structurally questionable to him, had in certain respects become inadequate, was in a state of transition. The axioms were only a matter of later formulation—after the real thing, the main discovery, had happened.[7]

[6] In our discussions Einstein focused on the material content of the steps. He did not use the terms of the preceding sentences of the text, terms which follow from the structural approach of this book.

[7] In this respect I wish to report some characteristic remarks of Einstein himself. Before the discovery that the crucial point, the solution, lay in the concept of time, more particularly in that of simultaneity, axioms played no role in the thought process—of this Einstein is sure. (The very moment he saw the gap, and realized the relevance of simultaneity, he knew this to be the crucial point for the solution.) But even afterward, in the final five weeks, it was not the axioms that came first. "No really productive man thinks in such a paper fashion," said Einstein. "The way the two triple sets of axioms are contrasted in the Einstein-Infeld book is not at all the way things happened

When we proceed with an analysis in the sense of traditional logic, we easily forget that actually all the operations were parts of a unitary and beautifully consistent picture, that they developed as parts within one line of thinking; that they arose, functioned, and had their meaning within the whole process as the situation, its structure, its needs and demands were faced. In trying to grasp the structure of this great line of thinking, the reader may find himself at a loss in view of the wealth of events, of the breadth of the situation. What, then, were the decisive steps?

Let us recapitulate briefly.

First there was what we may call the foreperiod. Einstein was puzzled by the question, first, of the velocity of light when the observer is in motion. He considered, secondly, the consequences as to the question of "absolute rest." Thirdly, he then tried to make one alternative workable (is the velocity of light in Maxwell's equations a variable?), and obtained a negative result. There was, fourth, the Michelson experiment which confirmed the other alternative—and, fifth, the Lorentz-Fitzgerald hypothesis, which did not seem to go to the root of the trouble.

So far everything, including the meaning and structural role of time, space, measurement, light, etc., was understood in terms of traditional physics—structure I.

In this troubled situation the question arose: Is this structure itself, in which the Michelson result seems contradictory, really clear to me? This was the revolutionary moment. Einstein felt that

---

in the process of actual thinking. This was merely a later formulation of the subject matter, just a question of how the thing could afterwards best be written. The axioms express essentials in a condensed form. Once one has found such things one enjoys formulating them in that way; but in this process they did not grow out of any manipulation of axioms."

He added, "These thoughts did not come in any verbal formulation. I very rarely think in words at all. A thought comes, and I may try to express it in words afterward." When I remarked that many report that their thinking is always in words, he only laughed. I once told Einstein of my impression that "direction" is an important factor in thought processes. To this he said, "Such things were very strongly present. During all those years there was a feeling of direction, of going straight toward something concrete. It is, of course, very hard to express that feeling in words; but it was decidedly the case, and clearly to be distinguished from later considerations about the rational form of the solution. Of course, behind such a direction there is always something logical; but I have it in a kind of survey, in a way visually."

the contradiction should be viewed without prejudice, that the time-honored structure should be requestioned. Was this structure I adequate? Was it clear just with regard to the critical point—the question of light in relation to the question of movement? Was it clear in the situation of the Michelson experiment? All these questions were asked in a passionate effort to understand. And then the procedure became more specific in one step after another.

How was the velocity of light to be measured in a moving system?

How was time to be measured under these circumstances?

What does simultaneity mean in such a system?

But, then, what does simultaneity mean if the term is referred to different places?

The meaning of simultaneity was clear if two events occur in the same place. But Einstein was suddenly struck by the fact that it was *not* equally clear for events in distant places. Here was a gap in any real understanding. He saw: It is blind simply to apply the customary meaning of simultaneity to these other cases. If simultaneity is to have a real meaning, we must raise the question of its factual recognition so that in concrete cases we can tell whether or not the term applies. (Clearly, this was a fundamental logical problem.)

The meaning of simultaneity in general had to be based on the clear simultaneity in the case of spatial coincidence. But this required that in every case of different location of two events the relative movement be taken into account. Thus the meaning, the structural role of simultaneity in its relation to movement underwent a radical change.

Immediately, corresponding requirements follow for the measurement of time in general, for the meaning, say, of a second, and for the measurement of space, since they must now depend upon relative movement. As a result, the concepts of time-flow, of space, and of the measurement both of time and space changed their meaning radically.

At this point the introduction of the observer and his system of co-ordinates seemed to introduce a fundamentally arbitrary or subjective factor. "But the reality," Einstein felt, "cannot be so arbitrary and subjective." In his desire to get rid of this arbitrary element and, at the same time, to get a concrete transformation formula between various systems, he realized that a basic invariant was needed, some factor that remains unaffected by the transition from one system to another. Obviously, both demands went in the same direction.

This led to the decisive step—the introduction of the velocity of light as the invariant. How would physics look if recentered with this as a starting point? Bold consequences followed one after another, and a new structure of physics was the consequence.

When Einstein reached the concrete transformation formula on the basis of this invariant, the Lorentz transformation appeared as a derivation—but now it was understood in a deeper, entirely new way, as a necessary formulation within the new structure of physics. The Michelson result, too, was now seen in an entirely new light, as a necessary result when the interplay of all relative measurements within the moving system was taken into account. Not the result was troublesome—he had felt that from the very beginning—but the behavior of the various items in the situation before finding the solution. With the deeper understanding of these items the result was required.

The picture was now improved. Einstein could proceed to the question of experimental verification.

In the briefest formulation: In a passionate desire for clearness, Einstein squarely faced the relation between the velocity of light and the movement of a system; confronted the theoretical structure of classical physics and the Michelson result.

A part-region in this field became crucial and was subjected to a radical examination.

Under this scrutiny a great gap was discovered (in the classical treatment of time).

The necessary steps for dealing with this difficulty were realized.

As a result, the meaning of all the items involved underwent a change.

When a last arbitrariness in the situation had been eliminated, a new structure of physics crystallized.

Plans were made to subject the new system to experimental test.

Radical structural changes were involved in the process, changes with regard to separateness and inner relatedness, grouping, centering, etc.; thereby deepening, changing the meaning of the items involved, their structural role, place, and function in the transition from structure I to structure II. It may be advisable to explain once more in what sense Einstein's achievement meant a change of structure.

1) In the Michelson situation—as in classical physics generally—time had been regarded as an independent variable and, therefore, as an independent tool in the business of measurement, entirely separate from, in no way functionally interdependent with the movements that were involved in that observational situation. Accordingly, the nature of time had been of no interest with regard to the apparently paradoxical result.

In Einstein's thought there arose an intimate relationship between time-values and the physical events themselves. Thus the role of time within the structure of physics was fundamentally altered.

This radical change was first clearly envisaged in the consideration of simultaneity. In a way, simultaneity split in two: the clear simultaneity of events in a given place and, related to it, but related by means of specific physical events, the simultaneity of events in different places, particularly under conditions of movement of the system.

2) As a consequence, space-values also changed their meaning and their role in the structure of physics. In the traditional view they, too, had been entirely separated from, independent of time and of physical events. Now an intimate relation was established. Space was no longer an empty and wholly indifferent container of physical facts. Space geometry became integrated with the dimension of time in a four-dimensional system, which in turn formed a new unitary structure with actual physical occurrences.

3) The velocity of light had so far been one velocity among many. Although the highest velocity known to the physicist, it had played the same role as other velocities. It had been fundamentally unrelated to the way in which time and space are measured. Now it was considered as closely bound up with time- and space-values, and as a fundamental fact in physics as a whole. Its role changed from that of a particular fact among many to that of a central issue in the system.

Many more items could be mentioned which changed their

meaning in the process, such as mass and energy, which now proved to be closely related. But it will not be necessary to discuss further particulars.

In appraising these transformations we must not forget that they took place in view of a gigantic given system. Every step had to be taken against a very strong gestalt—the traditional structure of physics, which fitted an enormous number of facts, apparently so flawless, so clear that any local change was bound to meet with the resistance of the whole strong and well-articulated structure. This was probably the reason why it took so long a time—seven years—until the crucial advance was made.

One could imagine that some of the necessary changes occurred to Einstein by chance, in a procedure of trial and error.[8] Scrutiny of Einstein's thought always showed that when a step was taken this happened because it was required. Quite generally, if one knows how Einstein thinks, one knows that any blind and fortuitous procedure is foreign to his mind.

The only point at which there could have been some doubt in this respect was the introduction of the constancy of light velocity in Einstein's general transformation formulas. In a thinker of lesser stature this could have happened through mere tentative generalization of the Lorentz formula. But actually the essential step was not reached in this fashion; there was no mathematical guesswork in it.

In late years Einstein often told me about the problems on which he was working at the time. There was never a blind step. When he dropped any direction, it was only because he realized that it would introduce ununderstandable, arbitrary factors. Sometimes it happened that Einstein was faced with the difficulty that the mathematical tools were not far enough developed to allow a real clarification; nonetheless he would not lose sight of his problem and would often succeed in finding a way eventually, in which the seemingly insuperable difficulties could be surmounted.

[8] In Act III, when Einstein examined whether a particular alternative would work, he actually did try several procedures. But although these attempts did not lead to a solution, they were by no means blind. At that stage it was wholly reasonable to test such possibilities.

CONCLUSION

Dynamics and Logic of Productive Thinking

I SHOULD love to go on telling of this exploratory trip, reporting further examples and the discussions to which they gave rise. But here I must come to a stop. I think that the few examples discussed may suffice for a first introduction. In these examples, in the concrete way of dealing with them, the reader will have seen some steps toward clarification, methods for deepening the problem, and the main lines of a new approach. Some points may be summarized briefly.

First, we have found what, in contrast to other processes, we may call genuine, fine, clean, direct, productive processes—better than some might have expected. It seems not to be true that men do not like or are generally unable to think that way. This is a result that may be deeply appreciated. Of course there are often strong external factors working against those processes as, e.g., blind habits, certain kinds of school drill, bias, or special interests.

Second, in these processes we found factors and operations at work—essential to thinking—which had not been realized by the traditional approaches, or had been neglected by them. The very nature of these operations, e.g., of grouping, of centering, of reorganization, etc., adequate to the structure of the situation (see Table III, pp. 190-191) is alien to the gist of the traditional approaches and to the operations which they consider.

Third, the features and operations described are of a characteristic nature: they are not piecemeal, they are related to whole-characteristics, they function with reference to such characteristics, determined by structural requirements for a sensible situation. In the context the items, data, relations, etc., arise and function as parts in their place and role within the whole, under the same dynamic requirements.

Fourth, while it is true that operations considered in the traditional interpretations are involved in the process (see Tables I, Ia and II, Introduction, pp. 6-9), they likewise function in relation to whole-characteristics. This is essential for the way in which they come into the picture.

Fifth, such processes are, on the whole, not of the character of an and-summative aggregation, of a succession of piecemeal, chance happenings in which items, associations, operations just occur. They are not arbitrary in nature: in spite of difficulties, in spite of some deviations and, often, of dramatic developments, thought processes show a consistency of development.

Sixth, in their development they often lead to sensible expectations, assumptions. These, as well as the other steps within the procedure, call for an honest attitude, for verification: in the absence of sincerity in one's attitude toward truth, there is the danger of dilettantism, of cheap plausibility. But the situation calls not merely for piecemeal factual truth, it calls for "structural truth."[1]

It is the features of points 2 to 6 that provide the concrete possibility for genuine, sensible, productive processes. (For other types of processes see p. 197 ff.).

Table III

Thinking consists in

envisaging, realizing structural features and structural requirements; proceeding in accordance with, and determined by, these requirements; thereby changing the situation in the direction of structural improvements, which involves:

that gaps, trouble-regions, disturbances, superficialities, etc., be viewed and dealt with structurally;

that inner structural relations—fitting or not fitting—be sought

[1] M. Wertheimer, "On Truth," *Social Research*, 1934, Vol. 1, pp. 135-146.

among such disturbances and the given situation as a whole and among its various parts;

that there be operations of structural grouping and segregation, of centering, etc.;

that operations be viewed and treated in their structural place, role, dynamic meaning, including realization of the changes which this involves;

realizing structural transposability, structural hierarchy, and separating structurally peripheral from fundamental features—a special case of grouping;

looking for structural rather than piecemeal truth.

In human terms there is at bottom the desire, the craving to face the true issue, the structural core, the radix of the situation; to go on from an unclear, inadequate relation to a clear, transparent, direct confrontation—straight from the heart of the thinker to the heart of his object, of his problem. All the items hold also for real attitudes and for action, just as they do for thinking processes. And thinking processes of this type themselves involve real attitudes.

Here again I have used such terms as "viewing," "looking for," "envisaging," etc., and I think they are appropriate and actually required. But many of the features in the table can, if desired, be expressed in objective or behavioral terms, as indicated in earlier chapters, i.e., by substituting "responses," "actions determined by structural features in the situation," etc.

To be sure, these terms are difficult. They are loaded, I think, with productive problems for research. Three groups of problems are involved in them which have to be envisaged and studied:

1. The features, laws, rules governing the neglected or barely investigated operations of segregation, grouping, centering, and structural transposability.

2. Problems as to the relation between parts and their wholes, etc., involving operations as to the place, role, function of a part in its whole.[2]

[2] Logistics has made contributions to these problems in a way neutral to 3.

3. Problems in regard to "outstanding wholes," the good gestalten, the ρ-relations.

Gestalt theory started scientific study of these problems in the search for theoretical clarification, for the laws involved, and tried to develop appropriate scientific tools for dealing with them in many experimental investigations. Without real knowledge of this literature the terms in the table are not easily understood, indeed they are easily misunderstood. It may suffice here that the reader regard the items as arrows pointing toward the concrete issues discussed in the various chapters.

Let us face the theoretical situation straight. Association theory, Approach II, and in many respects traditional logic, Approach I, show the following characteristics in their concrete operations, in the way they approach and center the picture:

In their aim to get at the elements of thinking they cut to pieces living thinking processes, deal with them blind to structure, assuming that the process is an aggregate, a sum of those elements. In dealing with processes of our type they can do nothing but dissect them, and thus show a dead picture stripped of all that is alive in them. Steps, operations come into the picture externally: on the basis of recall, of some previous knowledge, general or analogical, of associations in connection with some items in the situation (or even with the sum of them all), or again, of mere chance. The items, the connections used, are blind or neutral to questions of their specific structural function in the process. Such are the classical associations between an *a* and some *b*, the blind connections between means and end; such is the way in which traditional logic deals with propositions of the form "all S are *P*," or "if *A* then *B*." The connections, the items, data, operations are structure-blind or structure-neutral: blind to their structural dynamic function within the whole; and blind to the structural requirements.

All this makes direct grasp of productive processes of the type described impossible.

Dynamically, then, little more is given for theoretical understanding than the drive, the wish to get at the solution of a problem, and chance happenings, recall in terms of association, the assumption that what happened or what is true in many or in "all" cases will happen in this case too. Of course there is, besides, in traditional logic, the will to truth and to systematic knowledge.

The situation in the A-examples, given in the earlier chapters of this book (see Chap. I), clearly called for a theory of thinking that goes directly to the structural nature of these processes. It called for a theoretical approach in which what happens in such a process comes into the picture on the basis of vectors determined by the structural dynamics of the situation.

Generally speaking, there is first a situation

S_1, the situation in which the actual thought process starts, and then, after a number of steps,

S_2, in which the process ends, the problem is solved.

Let us consider the nature of situation 1 and situation 2 by comparing them, and let us then consider what goes on between, how and why. Clearly the process is a transition, a change from S_1 into S_2. S_1, as compared with S_2, is structurally incomplete, involves a gap or a structural trouble, whereas S_2 is in these respects structurally better, the gap is filled adequately, the structural trouble has disappeared; it is sensibly complete as against S_1.

When the problem is realized, S_1 contains structural strains and stresses that are resolved in S_2. The thesis is that the very character of the steps, of the operations, of the changes between S_1 and S_2 springs from the nature of the vectors set up in these structural troubles in the direction of helping the situation, of straightening it out structurally. This is quite in contrast to processes in which some steps, some operations coming from various sources and going in various directions, may lead to the solution in a fortuitous, zigzag way.

It is interesting to compare the psychological situation in cases in which, after the problem is put and seen, and the subject does not know how to proceed, someone arrives with the ready-made solution. The subject may or may not understand it, may or may not realize that this is the solution; in any case, it was not reached by him, it did not come into existence in the realization of the steps which are structurally required. It often gives him a shock, sometimes an unpleasant one. For real understanding one has to re-create the steps, the structural inner relatedness, the requiredness.

I repeat: The thesis is that the very structural features in S_1 with their particular, concrete nature create the vectors, in their direction, quality, intensity, that in turn lead to the steps and operations dynamically in line with the requirements. This development is determined by the so-called Prägnanz principle,[3] by the tendencies to the good gestalt, by the various gestalt laws.

Those particular cases appear here as the simplest archetype, in which S_1 is a structurally simple situation with no structural hiddenness, but with a structural gap or disturbance, and in which the change into S_2 is made by just getting things in shape. In such instances the structural requirements and the means to meet them are simply realized, and one often gets from almost all subjects a natural, easy, and forceful response. It is characteristic that these processes often occur even when no question is asked, no task is set: the problem itself arises in the structure of the given material.

In other cases, when the initial situation is not grasped either because it is too complex, too confusing, or because it appears in a simple but cheap, superficial structure, a transition is required first. The situation must be structurally understood so that the problem is grasped in its structural role as part of the given situation. Often this transformation actually explodes, revolutionizes the old view of S_1.

Here, stated briefly, is the gist of the thesis: structural reasons become causes in the process. Connected with the principle of sufficient reason, there was historically a long discussion concerning the relation of "reasons" and "causes." There was good ground for emphasizing that they are fundamentally different in nature. They are, no doubt, if understood in terms of Approach I. But the thesis here is that, with regard to structural reasons, the two coincide in sensible processes.

[3] Editors' note. The principle of Prägnanz, first formulated by Wertheimer with reference to perception, asserts that the organization of the field tends to be as simple and clear as the given conditions allow.

In other terms: When one grasps a problem situation, its structural features and requirements set up certain strains, stresses, tensions in the thinker. What happens in real thinking is that these strains and stresses are followed up, yield vectors in the direction of improvement of the situation, and change it accordingly. S_2 is a state of affairs that is held together by inner forces as a good structure in which there is harmony in the mutual requirements, and in which the parts are determined by the structure of the whole, as the whole is by the parts.

The process does not involve merely the given parts and their transformations. It works in conjunction with material that is structurally relevant but is selected from past experience, from previous knowledge and orientation.

In all this, such movements, steps, are strongly preferred as change the state of affairs in S_1 along a structurally consistent line into S_2.

If this is basically the nature of the process, i.e., if the steps are structurally determined, then various questions arise, for instance, why often the process does not proceed more directly, why there are states in which no progress is made, why the development may come to a dead stop and often remain blocked for some time—how deviations, mistakes originate. I have mentioned some of the reasons. I may repeat that a first inadequate view of the situation will often prevent the subject from grasping the real structure of the gap and the nature of the requirements that would enable him to close it adequately. Often the subject is lacking in breadth of view. Even when he has it at the beginning, he may lose it in the process because he is busy with details or falls into a piecemeal attitude. Under these circumstances closure may tend to occur in regions that are too narrow. On the other hand, of course, a subject's view may be overextended.

Often the possibility of a short-cut closure is seductive. When several part-problems are realized in a situation, the view of the whole may be lost as those partial views obtrude themselves. Often the impatient desire to find the solution focuses the eye too exclusively, too strongly, as a hungry animal separated from his food by bars, focuses on the near goal, loses the possibility of viewing the situation freely, and becomes unable to see that a simple detour would lead him to the goal.

And we must not forget that although the process $S_1 \ldots\ldots S_2$ is often a relatively closed whole, it is only relatively closed. It is a part-field, just as S_1 and S_2 are each only part of the field; and so is the whole process. It is a partial field within the general process of knowledge and insight, within the context of a broad historical development, within the social situation, and also within the subject's personal life. It is a part-field not entirely separated as to the material in the field, as to the amount and the sources of energy: conditions, factors, forces in the broader field, favorable or inimical, are of importance. Thus we have to take into account to what degree the partial field is separated, to what degree it is dynamically related to other parts in the broader field. But with regard tô this broader field, structural dynamics as outlined above for the partial field seem again to be the real issue. As a consequence, a wide field opens up for research and understanding in terms of inner structural dynamics.

Here I may mention one special point. The forces in the situation may be of two kinds. In many instances it is the structural nature of the objective situation which essentially determines the vectors and the steps, while the I, the ego, and his personal interests and tendencies play only a small role, or none at all. If concrete ego-tendencies come into the picture, they are often disturbing (cf. Chap. IV, section II). There are other cases in which personal needs are the source of the problem. Here the I plays an important role. But again (cf. Chap. IV, section I), in order really to solve the problem, a transformation is often needed first; the problem may remain insoluble so long as one focuses on one's own wish or need; it may become soluble only if, viewing one's desire as part in the situation, one realizes the objective structural requirements. In such cases one may reach the solution of one's goal sensibly, or else one may realize that the I-goal was itself blind and has to be sensibly changed, if not dropped altogether. So, even for the relation between the problem and the ego, structural features remain decisive.[4]

[4] See E. Levy, "Some aspects of the schizophrenic formal disturbance of thought," *Psychiatry*, 1943, Vol. VI, pp. 59-69.

So far I have limited the discussion to the case of $S_1 \ldots\ldots S_2$ as characterized above—the problem situation and the steps leading to its solution. I have mentioned, however, that often the process does not start with S_1 and end with S_2, but that in

$$\ldots\ldots S_1 \ldots\ldots S_2 \ldots\ldots,$$

S_1 already is part of a development; that, moreover, S_2, the very solution, does not represent an end but that by its nature it leads to further dynamic consequences. (See Chap. III, section V; Chap. V.)

There are other types. There is the type

$$S_1 \ldots\ldots$$

in which the situation in S_1 does not primarily involve a problem as to an S_2, a concrete goal. Rather, the real achievement consists in realizing that the situation is not in as good order as it looks; that it should be improved. Under these circumstances the process is often a transition from an and-sum or from a superficially structured view to a more adequate one. As a consequence, the first achievement would consist in realizing that there is a problem. To envisage, to put the right problem, is often a far more important achievement than to solve a set task.

On the other hand, there are processes in which S_1 plays little or no role. The process starts, as in some creative processes in art and music, by envisaging some features in an S_2 that is to be created. The artist is driven toward its crystallization, concretization, or full realization. Characteristically the more or less clearly conceived structural whole-qualities of the thing to be created are determining in the process. A composer does not usually put notes together in order to get some melody; he envisages the character of a melody *in statu nascendi* and proceeds from above as he tries to concretize it in all its parts.[5] For some composers this is not an easy process; often it takes a long time. When ideas about the goal are somewhat vague, colloidal, there may be two directions at work simultaneously—one working to get the central idea clearer, the other to get at the parts. Characteristically in such cases what would

[5] Similarly with a mathematician who envisages the idea of a formula or of an equation.

and what would not fit is immediately clear; whereas what happens in instances of the type $S_1 \ldots\ldots S_2$, is structurally determined by the nature of S_1, or of S_1 in relation to S_2, here determined by the structural features in the envisaged S_2, even though S_2 is still incomplete, still vague. This somewhat changes the dynamic nature of the outline given above, but in sensible procedure the vectors are again determined by the nature of the inner structural requirements.

Often two mutually related directions are present in the process, one which proceeds from some parts toward the whole, and another, which goes from whole-qualities to the parts. This is generally the case when a good gestalt is reached in a sensible process. Such a gestalt is not willfully imposed on parts irrespective of their nature; it answers their requirements too.

The dynamic theory sketched in these pages is not a smooth theory, does not mean to offer generalities just for the sake of subsumption, of cataloguing; it involves many real problems for research—marvelous problems, I think. But I do not think that the gist of it is foreign to what natural common sense feels.

I hope that the reader will not misunderstand the philosophical meaning of this approach. When a picture is given here of the inner structural dynamics in the determination of processes, it does not mean that in this development man is merely passive. An attitude is implied on his part, a willingness to face problems straight, a readiness to follow them up courageously and sincerely, a desire for improvement, in contrast with arbitrary, wilful, or slavish attitudes. This, I think, is one of the great attributes that constitute the dignity of man.

Central to the theory is the transition from piecemeal aggregation, superficial structure, to the objectively better or adequate structure. The criteria for the structurally true view are more difficult to establish than those of piecemeal truth. In this book I have confined myself to showing in comparatively elementary cases that exact decisions about the appropriate, the true structure can be made—as against skeptical, relativistic negativism.

Sometimes the situation is structurally ambiguous, as in ambiguous figures in perception when boundary lines may belong to one area or another, so that more than one possibility of structurization exists. So it is in many cases in which no particular structure is as yet the right one because our factual knowledge is too incomplete, and because data, facts, relevant to a decision are not at hand or are not established with sufficient clearness. Various conditions, forces, factors may determine a structure for the subject—factors which often include inertia of habits, piecemeal attitudes, and the working of the very Prägnanz tendency in the direction of premature closure. The subject then falls victim to a seductive simplification.

All this does not do away with the issue of objectively proper structurization. The desire not to be structurally blind, to get true structural orientation, seems strong, and it often shows itself even in the course and the fate of mistakes. To live in a fog, in an unsurveyable manifold of factors and forces that prevent a clear decision as to action, as to the main lines of the situation, is for many people an unbearable state of affairs. There is a tendency to structural clearness, surveyability, to truth as against petty views—the desire not to deceive oneself. If this desire to reach the true structure sincerely is weak, then structural simplification in a desired direction prevails. The extreme case is that of a paranoid system, in which data are misrepresented and actual facts are violated. Superficially miscentered structures are often dynamically precarious: Although progress may be slow because the forces in the structure make the subject overlook or avoid crucial issues, and rationalize his error, there are clear-cut cases in which an item or an argument endangers a superficial structural view, and explodes it in dramatically productive proceedings.

Such issues play an enormous role in the personal, social and political field. Often in political discussions, in political views, one realizes the impact of the Prägnanz principle in the almost irresistible tendency, the strong desire to get at a simple, decisive structurization of the field, to get clear-cut orientation, to act sensibly, not to be blind, not to act fortuitously. There is a thirst for true orientation.

In political discussions it often happens that not so much the facts themselves, the contents of arguments, are the issues in disagreement, but rather the structural role they play, the function they have in the context, with all the features of "because," "but," "nevertheless," "although," and so forth. Men are unhappy if the complication of such features befogs the issue; they long for a structurally clear view in which the items find their clear place, function and role, do not disturb the main lines and the resulting direction of view and action. This may lead them astray. But one often sees, too, how this tendency to structural simplicity is deeply connected with the thirst to get at the true structure. Experiences and experiments show this strongly and clearly in spite of forces which try to maintain an existing structural view.

In several experiments which I have made on these issues, there have been striking results of this kind. Dr. S. E. Asch is now likewise at work on a broad investigation of these issues, which were so much neglected by social psychologists because they focused almost exclusively on the study of arbitrary forces. I hope that Dr. Asch will soon publish his findings.

Here lie great tasks for democracy. Critical attitudes, skepticism, do not suffice. What is needed is structural clearness. There is the hope that productive methods will be improved, not merely for gathering information about piecemeal facts, but also for gaining clear insight into the great lines, the basic structures of crucial situations.

Now, besides such processes as are discussed in the chapters of this book (type α), there are many others which to a greater or lesser degree contain features of another nature (type β). Even in processes of the kind described, some of the items needed for progress, or some of the operations, may drop in externally, by chance, by external analogy, by mere recall, or as a result of blind trying. Besides, on the borderline of the known, in developing science, there are too many situations the nature of which calls in the first instance for careful investigation of facts, for realization of factual relations, etc., because there is as yet too little known, too little understood. But those are marvelous moments when, after a long period of diligent, careful study or experimentation, a way opens for structural understanding; or when an experiment brings results which do not fit, which even contradict a given structural view, and the process then proceeds under this challenge.

The extreme at the other end (γ) are cases in which the result, the solution, is brought about by sheer chance discovery or merely by a succession of blind trials, by sheer external recall, sheer reliance on blind repetition, by blind drill or by prompting. There are many situations the nature of which fundamentally allows of nothing but blind proceeding and blind finding as, for instance, in widely used experiments with mazes, discrimination tasks, and problem boxes. Here all the factors that might furnish some clue to reasonably directed behavior are carefully excluded by the experimenter. Under these conditions no genius, however great, could at first do anything but engage in blind trials; success could occur only by chance, and then be repeated—unless meanwhile the arbitrary set is changed arbitrarily by the experimenter.

To repeat: The differences between the extreme α and γ concern not merely intellectual procedures; they involve deep differences in human attitude.

Many theorists try to center the whole picture of thinking around the characterization of type β, and thus neglect the structural features which the β processes, too, contain.

In modern psychology there is a strong trend to regard thinking basically in terms of the factors, operations, and attitudes of type γ, to be blind to the possibilities of type α, and to try by all means to interpret the types β and α as nothing but complications of factors characteristic of type γ. The study of such factors is undoubtedly needed. But one should not generalize too easily, too superficially. Even in cases in which it is possible to construct a piecemeal, blind mechanism for the "explanation" of a process, it is appropriate for the scientist to be careful lest instead of a true picture he may get a poor, only superficially adequate substitute. In this respect one should also be particularly careful because of the serious implications involved for questions of teaching and education, of life.

The situation is similar to the situation in the psychology of learning.[6] Type γ corresponds to learning by drill, by external associations, by external conditioning, by memorizing, by blind trial and error.[7] Type α focuses on developing structural insight, structural mastery, and meaningful learning in the real sense of the word. There is a widespread assumption that sensible learning, that learning of sensible material, is at bottom nothing but a complication of what is found in the rote learning of syllable series, etc., as though this latter provided *the* laws of learning. It does not seem possible that the characteristics in α are reducible to factors and operations of this kind. Even if one should entertain such a hope, it is not founded in real investigation but often merely plays the role of a dogma.

To put it in the briefest way: If we call the thinking and the learning processes of type α "structurally sensible," if we call the characteristics of type γ "structurally blind," the situation in this traditional view is:

FIG. 147

In other words, γ is taken as basic, α "will certainly turn out to be nothing but a complication of γ factors."

The scientifically more careful procedure is first to study the distinctive character of each type of process. Only on the basis of

FIG. 148

such studies will it be possible to decide whether the two types are entirely different in nature, or whether α is to be viewed as a com-

[6] See my introduction to G. Katona, *Organizing and Memorizing* (Columbia University Press, 1940).

[7] Editors' note. By "external" association Professor Wertheimer meant connections in memory which are supposed to be established irrespective of the content of the items involved. The term "external" conditioning is to be understood in the same way.

plication of the factors essential in γ, *or* whether α is the proper theoretical center and γ a special case.

Fig. 149

At the present time the last seems to be the theoretically more appropriate alternative: γ appears to be merely a special case in which the structural interdependence characteristic of type α approaches zero, a limit which is never reached in cases of real learning, of real thinking.

Let us relate this to a previous distinction among several approaches, which has occupied us before. If we look back to Table III (the gestalt approach) and compare it with the approaches of traditional logic (both deductive and inductive) and of association theory (see Tables I, Ia, and II, Chap. I, pp. 6-9), two ways are open to us. Either we regard the structural characterization of III as complicating additions to I and II, or we decide what an appropriate theoretical attitude may be only after the functional principles of these approaches and their mutual relations have actually been studied. To be sure, each of the items in I and II is important. But perhaps these operations themselves merely represent special cases. The items in II and to a degree in I were conceived and traditionally used as neutral or blind to structural features and requirements. Scrutinizing them, we find that each of the items in I and II is itself ambiguous, that each can be meant in a structurally sensible or a structurally blind way. Their structurally blind form and operation seems to be a limiting case of III, which is adequate only when structural coherence and interdependence approach zero.

This does not mean that regions possessing the appearance of I and II, and therefore lacking structural characteristics as to content and connections, are *altogether* bare of structural factors. Even if connections are merely factual, merely factually constant, and not understandable, the *hierarchy* of such connections still offers possibilities either of structurally sensible or structurally blind proceeding.

Let me now briefly characterize the ambiguity of the terms and operations of Tables I, Ia, and II.

The Terms of Traditional Deductive Logic: Table I (Introduction, p. 6).

Comparison and *discrimination* may mean, and are commonly supposed to mean, that two or many objects are compared with regard to just any features in them, blind to the given structure. From this point of view what matters is solely whether or not there are equalities or differences, and what they are. But equalities may mean piecemeal sameness which may mislead even if they are constant and general; or, quite in contrast, there may be structural equality, which can obtain even when the piecemeal data show no equality at all.

Analysis may mean that a field or an object is cut into and-summative parts, blind to structure, or it may mean structurally adequate division, and viewing parts in their part-nature.

Abstraction and *generalization* may mean a performance that focuses on pieces, is blind to structure, and leads to the and-summative form

$$m + x.$$

Here m represents facts common to several situations, and x other characteristics in which these situations differ (cf. p. 207). Now the existence of the common factor may mean no more than the coincidence of certain pieces or qualities, stated without any regard to their role in the given structure. The procedure may even involve divisions and cuts which violate their structure. On the other hand, abstraction and generalization may also mean operations that follow the requirements of given structures. Similarly with *class* concepts. The grouping into classes and subclasses may be done in such a way that it groups together objects and classes which are structurally foreign to each other, therefore basically different, and sharply separates objects which are structurally similar or even structurally identical (see pp. 208-210). Conversely, class concepts may refer precisely to those common structural factors which the first procedure ignores.

Propositions, for instance, of the character "all S are P," may state a factual and rigid but blind connection, a factual coexistence of facts that structurally do not in the least belong to each other; or again, they may be structurally sensible predications. A set of predicates attributed to a subject may either mean an and-sum of unstructured data; or it may refer to data that fit one another, and it may thus illuminate the very structure of S.

Similarly with *inferences, syllogisms*, etc. They may be conceived and used in terms of merely formal relations in which such empty quantifications as "all," "some," "none" play the essential role; or they may spring from structural requirements.[8]

The Terms of Traditional Inductive Logic: Table Ia (Introduction, p. 8).

Induction may mean generalization on the basis of piecemeal, external coincidences in a number of cases; or it may mean a structurally reasonable hypothesis.

Experience may mean a heedless gathering of facts and of mere factual connections. Or experience may mean that structural features are vividly grasped, that orientation is gained, that data and connections are understood in their role and function within the context.

Experimentation may mean that just any piecemeal factors are heedlessly introduced and that results are viewed piecemeal, without regard to their structural meaning. This is often necessary as a first step. But more is needed lest in the end there be nothing but a sheer and-sum of structurally unrelated facts. At the other extreme lies structurally sensible experimenting, often in the form of asking nature a crucial question, of trying to decide between possible hypotheses in the structural context of knowledge.

"One variable is a function of another variable." This may mean, as some theorists consistently formulated it, that items of any two

[8] See the paper on syllogisms in productive thinking in which the empty though exact syllogisms were contrasted with sensible syllogisms. M. Wertheimer, "Über Schlussprozesse im produktiven Denken," *Drei Abhandlungen zur Gestalttheorie* (Erlangen, 1925) pp. 164-184; also W. D. Ellis, *op. cit.*, selection 23.

series of facts are correlated, and that a principle is given in terms of correlated changes without regard to the structural sense of the pairing. With this concept of function one does not consider how the pairing and the principle are related to the nature of the items and to the structural characteristics of the whole series. At the opposite extreme one may study what changes in a part may mean within the structure of the whole; one may thereby discover the inner laws governing the nature of the items within the whole, the way in which their changes depend upon this part-whole relation.

The Terms of Association Theory: Table II (Introduction, p. 9).

Association may mean chaining items together in an and-sum of connections that has by nature no structure, as in the usual theory of learning rote syllables. Or at the opposite extreme, it may mean the realizing of structural belonging in which items require each other as parts in a context—including the enduring effects of that realization.

Repetition may mean that the same piecemeal, blind connection occurs over and over again; or it may mean the change from an ununderstood and sheer additive pairing to the realization of a structure in which the meaning of the items becomes that of parts in a characteristic whole.

Trial and error may mean a heedless succession of blind proceedings with random order of directions; or, again, it may mean that some sensible hypothesis is structurally tested. In the latter case, the very failure may elucidate the situation and suggest another hypothesis which fits the given structure better.

Learning on the basis of success may mean that an action is singled out because of the success that follows the action only factually, but is not understood; or it may mean that, in learning, a subject grasps why just this kind of action leads to just this effect for intrinsic structural reasons. It is the latter form of "learning by success" that enables the subject to vary his action in a structurally sensible way when the situation is no longer the same.

The essential difference between the two interpretations of all these concepts can perhaps best be illustrated if we once more return to logic and, in the first place, to the class concept which is so fundamental in the traditional form of that discipline. If we disregard the many intricacies, if we concentrate on just the concrete meaning of the operations involved and on what is really needed for correctness in traditional logic, we find the following points:

There are several objects. (The way in which they are segregated, and why just so, how an object constitutes itself in separation from other objects, is a question neglected in traditional logic, is taken for granted without real investigation.) I compare them. In their qualities or their parts I find similarities and differences. Abstracting from the differences, and concentrating on common qualities or parts in the objects, I get a general concept. The content is given by these common parts. This is the "intension." The "extension" is the manifold of objects embraced by the class concept.

If we call the common element *m*, and the other elements *x*, an exact expression for the class (or for any object as conceived under the class concept) is

$$m + x.$$

Between the *m* and the *x* is an "and." The *m*[9] is what is common in the contents of the objects; the *x* is what there is besides the *m* and may vary in the contents of the various objects. The conceived datum, *m*, is independent of its setting to the left and right, and apparently must be so for the sake of exact use of the concept in inference, syllogisms, etc. There is no reference to whatever else there may be in the object besides, no references to the role which *m* plays in this object, no reference to its meaning as a part among the other parts of the same entity, no reference to the structure of this entity. This abstraction is subtractive; it simply isolates the *m*. For the *m* it does not matter what the *x* is. The *x* is in principle arbitrary; in other words, no question is raised as to what the *x* may be, and what the *x* may mean for the *m*. The rigid constancy of the *m*, and the independence of its nature from the nature of the *x*, is assumed to be basically needed for the proper performance of classification, subsumption, universal propositions, inferences, syllogisms, etc. as envisaged by traditional logic.

[9] The *m* itself may be an and-sum of several common elements.

In many cases such a procedure is adequate and useful, as in often used classical examples of traditional logic; consider the proposition "All mailboxes in the . . . State are green." It is fully adequate in all cases in which the *m* and the *x* are just separate, additive, merely put together without any inner relation that would make them mutually dependent; in all cases in which the meaning of the *m* is invariant with regard to changes of the *x*, or vice versa.

In the historic development difficulties have arisen as to the adequacy of the procedure in certain cases (cf. the famous discussion of the Linnaean system of plants in the French Academy). The problem was whether such a procedure, although exact, does not easily combine objects which are basically different in nature and, on the other hand, sharply separate objects which belong to each other in fact. The logician seeks help in the term "essential." There always was emphasis on this point; but although for common sense the meaning of "essential" is often clear enough, unfortunately it was and has remained extremely controversial in logic. It has served to name the problem rather than to solve it. It has consequently been rejected again, excluded in newer developments of logic. A way to clarify its beautiful meaning opens up when we attend to structural features. I give an extreme example from music. Here are four objects:

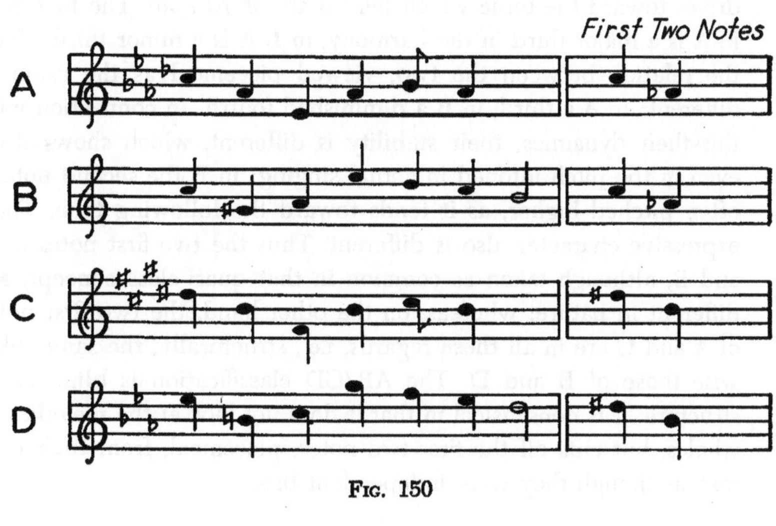

FIG. 150

We classify. We realize, with regard to the pitches, that the objects A and B start with the same two notes. So do the objects C and D. A librarian may form a first class of melodies that start with the first two notes of A and B, and a second class that start with the first two notes of C and D. This may—although I wonder—be useful to him for bringing some exact order into his collection index. In terms of traditional logic the procedure is exact. But what would he have done in this procedure? He would have combined the first two melodies which are different in nature even as to these two first notes. And so in the second class. He would have sharply separated into different classes melodies which are the same, only in another key, C being a transposition of A, D of B.

On the piano the first two notes in his classification are the same in A and B, C and D, respectively, but they are not the same for somebody who grasps the melodies. For him the two notes which the classification, with its atomistic procedure, regards as identical are actually very different as to the role they play in the melody, different also as parts of it. If one were to write these "identical" notes with the same symbols—as I have done on page 208—the musician would be exasperated, would call it nonsense, illogical. The second note in A is the tonic, the "identical" second note in B is far from being a tonic, it is the leading tone which asks for, drives toward the tonic which here is the third note. The first note in A is a major third in the harmony, in B it is a minor third. Even the relation between the two, viewed piecemeal as the same, is different: in A a third, in B a diminished fourth. In connection with this their dynamics, their stability is different, which shows itself even in the pitch intoned in actual singing: in B the second note is often pitched higher, as it tends toward the following tone. Their expressive character also is different. Thus the two first notes of A and B, although taken as common in that quasi-class concept, are different in nature; whereas, on the other hand, the two first notes of A and C are in all these regards, i.e., structurally, the same, likewise those of B and D. The AB/CD classification is blind as to structure; it is nonsensical in that it does not look at the melodies as wholes, but cuts off the first two notes, piecemeal, from their context, as though they were independent bits.

Examine the contrast: The structure-blind class formation gives the grouping AB/CD, the structural formation gives the grouping AC/BD.

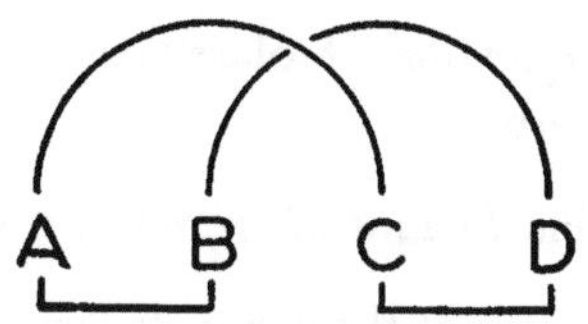

FIG. 151

Here we have considered only strict transposition; but in sensible musical variations even the two starting notes of a melody and their interval may to a certain extent be changed with no real damage to the melody as a structure. On the other hand, a single note, when changed, may be out of place, may violate the structure. When one grasps such a melody one realizes that something is wrong, is out of shape, does not fit. Melodies which are violated in such a way, and senseless tone-aggregations generally, do not transpose psychologically as good melodies do. When there are structural violations, there is trouble. If one tries to recall senseless tone aggregations, to repeat them after a time, they are much harder actually to repeat, and various things may happen to them. There seems to be a strong tendency to change, somehow to improve such material in the direction of a sensible structure. So, far from being a matter of piecemeal equalities, the real issue is not even restricted to questions of place, role, and function in some whole; but it refers to fitting or not fitting given structural requirements.

I have chosen the example of melodies because in musical grasping the issues are vividly felt. Of course exact formulation of the whole-qualities, the structural requirements, is not easy here—it is what some great musicians have called the inner logic of melodies, one of the great problems in aesthetics. Yet much of what I have tried to show in these examples is of general import; it holds often for different materials, when exact formulation in these respects offers no difficulty. The same issues can, for instance, be studied in groups of, say, four objects in figural distributions, in the structure of events within physical systems, in abstract relational networks, and in sets of human traits. A broad field opens up, much broader than the issue of classification, when we consider the problems of transposability and search for the principles of structural invariance.

In regard to the question of classification the main point boils down to the old proverb,

si duo faciunt idem, non est idem,

if two do the same, it is not the same. In exact terms: Two items or two groups of items which are identical from an atomistic point of view (cf. above AB/CD), may mean very different things structurally, may in fact be different in nature. A necessary addition is the opposite sentence: If, from an atomistic point of view, two do very different things (cf. above AC/BD), their actions may nevertheless be structurally the same. In order to do the same in a changed situation one has to do it differently. In exact terms: Different items may be structurally the same.

Similar things hold for the items which are usually considered to be the most basic ones in logic: for the "and," the "not," the "if-so," the concept of relation, of identity, of truth, etc. I shall mention some points briefly. Traditionally all of them have been considered and used in blindness to the structural problems. All of them, in their traditional meaning, seem to be merely limiting cases of the broader issue. This holds even for the traditional principles of thought: the law of identity, of contradiction, the principle of sufficient reason.

In exact traditional logic "*and*" may combine any two things, or propositions, whatever they may mean to each other, whether or not they belong to each other structurally. "And" then means the one *is*, or *is true*, also the other. I use a characteristic example from the classical treatise of D. Hilbert and W. Ackermann.[10] One can exemplify the meaning of the traditional "and" by a statement such as, "Two is smaller than three *and* the snow is white." Here we see that the content of both, taken together, is nothing but their and-sum; the actual content of each means nothing to the actual content of the other; the subject matter of the two has no inner structural

[10] *Grundzuege der theoretischen Logik* (J. Springer, Berlin, 1928) p. 3.

relation. In the and-sum each is just what it would be without the other, or with the other changed. This example may shock the reader, but it gives the exact meaning of the "and" in a structure-blind logic.

In fact, this empty "and" is merely an extreme case. In living thought "and" is for the most part not of this character. There is the "and" that combines two things which belong to each other, which require each other structurally. There is the "and" that states two things are together which should not be together, which violate each other. Both these are different functionally from the neutral, structure-blind "and." The real "and" is often of a very serious character in that it involves dynamic consequences to which the use of the empty "and" is blind. Even in formal logic we should differentiate basically between the various kinds of "and," because the general use of the empty "and" may blind the thinker to what he really does in putting things together.

In modern logistics the "and" has been defined by a set of truth-values as to two propositions. Elegant in itself, the procedure strikingly expresses an underlying structure-blindness with regard to the "and," and to the meaning of the two propositions. It is adequate for cases in which the two propositions deal with subject matters that have no structural relation to each other, in regard to which "and" properly means nothing except that each is true independently. But there are cases in which the combination of two subject matters is not of a sheer "and"-summative character. If we then consider first each proposition alone, separately, and afterward realize what happens when they are related by a real "and," we shall find that this often brings about serious changes in their meanings. To some extent this can already be shown with great exactness in regard to propositions on simple relational networks—when the changes occur in terms of the so-called implicit definitions of the logistician, and consequently with regard to the table of the truth-values itself. The form of the table of the truth-values in the mere and-summative case appears merely as a limiting case of the very much richer and deeper set which we get by considering structural features.

To summarize: The real "and" implies real relations, the existence of specific wholes and of their dynamics.

It is a serious matter, characteristic for the structure-blindness and function-blindness of traditional logic, that such terms as "but," "nevertheless," "however," were not considered in the logical system at all.

Similar things hold for the meaning of a "*relation*" as for the "and." There are cases, in which

$$|a|\ |R|\ |b|$$

constitutes a mere and-sum in which none of the three items really means anything to the others. A relation, *R*, to a *b* is stated for an *a* with no consequence at all for the *a*, and none for the *b*. Secondly, there are cases in which two things or two items are put into some real relation to each other, which is structurally inadequate to both, violates the requirements of each and yet they have to live it. This is the form of relationship in which violent structural dynamics often result. Thirdly, however, there are cases in which in the actual relation, the items complete each other in a good structure, fit together, and build a good whole.

Eventually there are the cases in which the items mutually determine one another from inner necessity, and in a clear, clean way; where *a* and *b* determine *c*, or require their *R*; where *a* and *R* ask for the adequate *b*, and *R* and *b* for the adequate *a*.

Just as with "and" and with "relation," we find that the concept of negation may be taken in an empty, structure-blind sense. But, again, this is a limiting case of negation, applicable only in special instances. On the other hand, negating something may mean that this something is appropriately not the case, the negation being just what is required by the structural nature of the situation. But there is still another "not," which states the lack of something that would in fact be structurally required by the situation. Both are in contrast to the case of an empty negation, which has no structural meaning whatsoever. The case of the negation that states a structural lack is actually the *negatio privativa* of classical logic; but it is essential that its structural nature be clearly understood. Between the empty negation and the other forms of "not" there lies a manifold of forms.

Similar distinctions hold for the *"if-so"* form which is so fundamental in logic. The one extreme is the structure-blind, merely formal *"if* two is smaller than three, *then* snow is white."[11] For formal purposes it is important to study this most empty, structure-blind type also. We sometimes have to deal with constant connections of this kind in actual life, sometimes even in the initial phases of productive processes. But in sensible thinking "if-so" is rarely, or at least almost never *remains,* of this empty kind. Common sense is right in feeling shocked by such examples as the one just given. "If-so" for the most part involves some structural justification. It does not simply mean an "if-so" with regard to structurally unrelated subject matter. The sensible "if-so" calls for some kind of inner coherence, some kind of intrinsic structural relatedness. Thus the empty type appears merely as an extreme in which all structural relatedness disappears and only an external form remains, which is indifferent to the issues under "if" and under "so."

Or take the principle of identity. The case of full identity is banal and, in actual thinking, hardly ever an issue. The real problem refers to the discovery of "identity" in spite of some apparent differences, and here it becomes a fundamental task to discriminate between piecemeal identity that ignores structure, and structural identity. It has been possible to study these differences in psychological experiments. The investigation, which referred to concrete subject matter, led to the clear-cut consequence that piecemeal identity is merely a special case, appropriate only when structural conditions allow.[12]

It is the same with *truth* itself. Study of the problem of truth gave rise (see the paper quoted p. 139) to the scheme of a four-valuate logic with the terms: true or false, each to be taken either in an atomistic or in a structural sense.[13] Here again structure-blind pro-

[11] This example is not my invention. It has been used for positive characterization (D. Hilbert and W. Ackermann, *op. cit.*, p. 4).

[12] J. Ternus, "Experimentelle Untersuchungen über phänomenale Identität," *Psychologische Forschung*, 1926, Vol. 7, pp. 81-136.

[13] M. Wertheimer. "On Truth," *Social Research*, 1934, Vol. 1, pp. 135-146.

cedure leads to the special case of the merely two-valued Aristotelian logic.

All these issues play an important role in productive thinking. But in this connection they ought to be considered as parts of the great issue of the *dynamics* of thinking. Whereas traditional logic focused on the problems of validity, on static features, the doctrine of general logic has to deal with the logical features and rules of dynamic events, and these again are structural.

For instance, our statement that identity has often to be understood in a structural sense does not entirely suffice. Traditional logic regards it as a very basic rule that the items of discourse—concepts, propositions, and so on—have to remain rigidly identical if repeated. Important as this rule is for certain questions of validity, it does not generally fit real thinking. In real thinking processes, items often do not remain rigidly identical; and as a matter of fact, precisely their change, their improvement is required. If an item, concept or proposition, recurs in the process and appears from an atomistic point of view as identical, it very often is not really so. Its functional and structural meaning has actually, and fortunately, changed. Blindness to such a change in meaning often impedes productive processes. In real thinking the *functional* meaning of an item, of a proposition, that meaning which changes as thinking advances, is of the utmost importance—without it thinking gets sterile; without realization of that change one does not grasp the line of progress. For statements, etc., have a *direction* in their context. It is here that a basic feature of traditional logic comes to the fore: its disregard of the intense directedness of live thought processes as they improve a given situation.

APPENDIX

The Sum of a Series

I SHALL describe how a direct train of thought proceeded.[1] I shall express the series in the form

$$\frac{1}{a} + \frac{1}{a^2} + \frac{1}{a^3} + \cdots \qquad (a \text{ is an integer} > 1)$$

The problem is to find the sum of the continuous series.

First: How does the sum increase if the series continues? Obviously the sum in each successive step, S_{n+1}, is equal to the previous S_n plus $\frac{1}{a^{n+1}}$. The value of $\frac{1}{a^{n+1}}$ becomes smaller and smaller the farther I proceed in the series: the larger n becomes, the smaller will be the value $\frac{1}{a^{n+1}}$. Hence in later stages S_{n+1} will more and more approximate S_n—the curve of the sum will no longer rise so much, will change less and less, and will perhaps approach the horizontal.

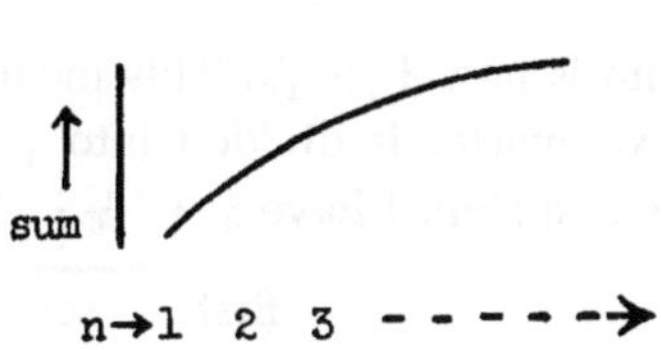

FIG. 152

But this does not help me to see what value the sum is approaching.

[1] This report was written down during the actual thought process.

Second: If I want to understand, I must realize from the beginning what the first term, $\frac{1}{a}$, means as part of its whole; and so with the later terms.

$\frac{1}{a}$ means that 1 is divided into a equal parts and that $\frac{1}{a}$ is one of them; e.g., $\frac{1}{4}$ means:

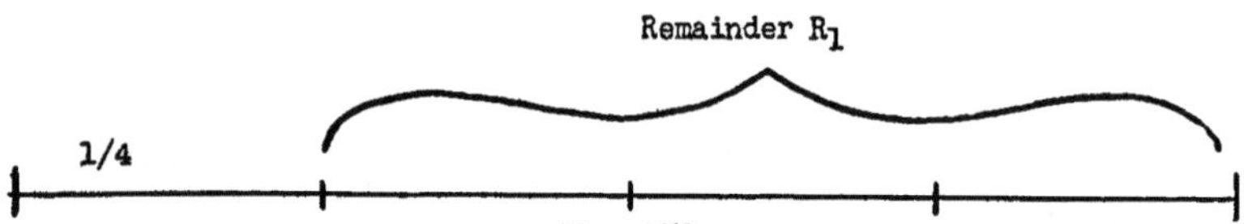

FIG. 153

I have divided 1, the whole, into 4 equal parts; in considering the first of these parts, $\frac{1}{4}$, I have to realize that the other parts remain, in the present case, $\frac{3}{4}$. In general, taking $\frac{1}{a}$, the whole 1 is divided into $\frac{1}{a} + (a-1)\frac{1}{a}$. This is the situation with the first member of my series.

What happens if I proceed to the next step, $S_2 = \frac{1}{a} + \frac{1}{a^2}$?

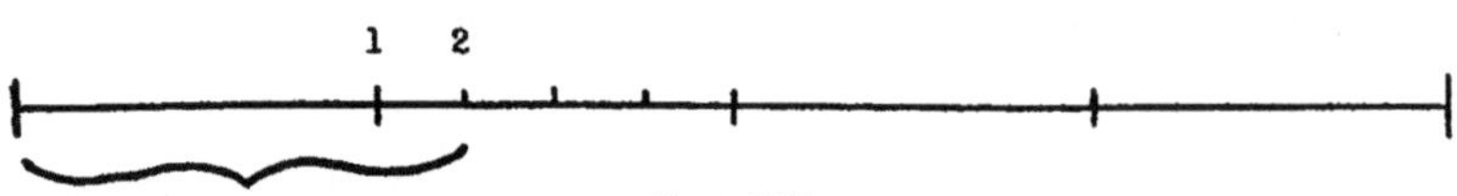

FIG. 154

In our case the sum is now $\frac{1}{4} + \frac{1}{16}$. This means that with the addition of $\frac{1}{16}$, the next quarter is divided into $\frac{1}{16} + \frac{3}{16}$, as the whole was divided in the first step. I have

$$\underset{\text{first quarter}}{\frac{1}{4}} + \underbrace{\frac{1}{16} \vdots + \frac{3}{16}}_{\text{second quarter}} + \underset{\text{remainder } R_2}{\frac{2}{4}}$$

Or in general,

$$\underset{\text{first } a \text{ part}}{\frac{1}{a}} + \underbrace{\frac{1}{a^2} \vdots + \frac{a-1}{a^2}}_{\text{second } a \text{ part}} + \underset{\text{remainder } R_2}{\frac{a-2}{a}}.$$

R_2 must now be $(a-2)\frac{1}{a}$ because the first part of R_1 has been drawn upon in the $\frac{1}{a^2}$ division.

Again if I proceed to the next sum, $\frac{1}{a} + \frac{1}{a^2} + \frac{1}{a^3}$, the following happens:

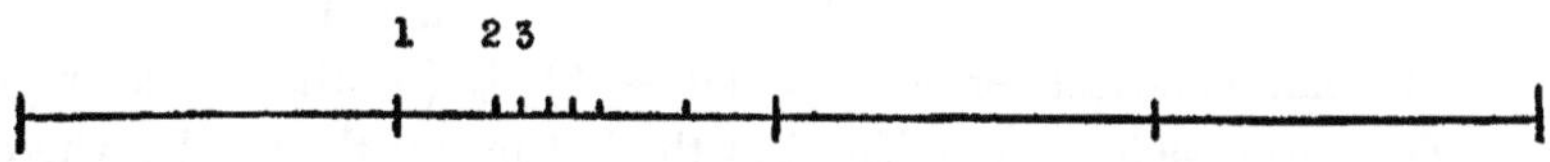

FIG. 155

Here the next $\frac{1}{16}$ is divided into $\frac{1}{64} + \frac{3}{64}$; I have

$$\frac{1}{4} + \underbrace{\frac{1}{16} + \underbrace{\frac{1}{64} \;\vdots\; + \frac{3}{64}}_{\text{second } \frac{1}{16}} + \frac{2}{16}}_{\text{second } \frac{1}{4}} + \frac{2}{4}$$

Or in general, $$\frac{1}{a} + \frac{1}{a^2} + \overbrace{\frac{1}{a^3} \;\vdots\; + \frac{a-1}{a^3}}^{\text{critical region}} + \frac{a-2}{a^2} + \frac{a-2}{a}.$$

Every time, in every step, the next region $\frac{1}{a^n}$ is divided into $\frac{1}{a^{n+1}} + \frac{a-1}{a^{n+1}}$; and with that the previous $\frac{a-1}{a^n}$ changes into $\frac{a-2}{a^n}$. In this way, the critical region $\frac{1}{a^{n+1}} + \frac{a-1}{a^{n+1}}$ shifts slowly to the right, and decreases in size at the same time; the structure is

$$S_n + \overbrace{\frac{1}{a^{n+1}} + \frac{a-1}{a^{n+1}}}^{\text{critical region}} + (a-2)\,S_n$$

R_n in the general structure must be $(a-2)$ times S_n itself—it changes from $(a-2)\,\frac{1}{a}$ to $(a-2)\,(\frac{1}{a} + \frac{1}{a^2})$, to $(a-2)\,(\frac{1}{a} + \frac{1}{a^2} + \frac{1}{a^3})$, and so on.

Third: The whole, "1", is divided into S_n, the critical region and $(a-2)\,S_n$, the remainder R_n. As we proceed, the critical region becomes necessarily smaller and smaller, while the sum at the left increases. If the critical region becomes exceedingly small, approaches zero, the whole appears to be divided into $S_n + (a-2)\,S_n$:

the whole no longer appears to be divided into *a* equal parts, as in the beginning, but into $(a - 1)$ *equal parts*! So the sum of the continuous series itself must approach the value $\frac{1}{a-1}$!

In mathematical terms: $1 = (a - 1)\ S_n$ (+ the critical region which approaches zero). The 1 at the left means the whole, whereas in the usual derivation it is just the first member of the series $1 + a + a^2 + \cdots\cdots$; here there is the realization of what S is structurally as part of its whole, whereas in the usual derivation the value of S is reached by some algebraic operations.

Fourth: I want to formulate exactly. I had

$$1 = S_n + \text{critical region} + (a - 2)\ S_n$$

The value of the critical region is $\frac{1}{a^{n+1}} + \frac{a-1}{a^{n+1}} = \frac{a}{a^{n+1}} = \frac{1}{a^n}$.

As we saw before, it is the next $\frac{1}{a^n}$ which becomes divided in the transition from S_n to S_{n+1}. Therefore, in exact formulation

$$1 = (a - 1)\ S_n + \frac{1}{a^n}$$

$$\text{or,} \qquad S_n = \frac{1}{a-1} - \frac{1}{a^n(a-1)}.$$

$\frac{1}{a^n(a-1)}$ decreases as *n* increases; it never reaches, but approaches zero more and more closely.

Fifth: The whole procedure consisted in realizing the meaning of the parts in the whole structure, which led to structurization of the whole into $(a - 1)$ parts plus the diminishing critical region.

The result clearly holds for whatever value $\frac{1}{a}$ may have, provided *a* is greater than 1. In the series starting with $\frac{1}{2}$, of course there are no $(a - 2)$ parts; $a - 2$ being zero, the series approaches 1. If $\frac{1}{a}$ is equal to or greater than 1, the conditions for this structurization are not given—the sum will forever increase, will not approach a limit.

But what of values between $\frac{1}{2}$ and 1, e.g., $\frac{3}{4}$? There seemed at first to be no way to structural realization. The way was opened by viewing $\frac{3}{4}$ not as part of 1 but as $\frac{1}{4}$ of 3, taking 3 as "the whole":

FIG. 156

This gave the basis for understanding the fate of the critical region, etc., which is here in an interesting way more complex than in the preceding cases. But I shall not report the further steps dealing with this and other related problems.

If we look back to the usual proceeding—multiplication of the series and subtraction—this becomes clear: Although the usual procedure is so very much quicker, externally so much more elegant, though it holds equally generally for the structures first considered and for the more complex instances, it is nevertheless a trick in the sense that it gives no direct understanding of what happens structurally.

Of course such elegant external procedures, leading to quick solutions, are useful. But the longer way is needed in order really to understand the construction, to realize what happens. As for *finding* the solution, the first way is either a chance happening only, or a clever trick, the result of some valid equations, which bring the solution without giving insight into the nature of the series. To be sure, it may have started with structural insight into *one* feature of the series, viz., that it is shiftable, that if one multiplies by a or a^2, the same series appears except for the first members; but this is not sufficient to reach the solution by way of structural understanding.

Index

MIX
Papier aus verantwortungsvollen Quellen
Paper from responsible sources
FSC® C105338

If you have any concerns about our products,
you can contact us on
ProductSafety@springernature.com

In case Publisher is established outside the EU,
the EU authorized representative is:
Springer Nature Customer Service Center GmbH
Europaplatz 3, 69115 Heidelberg, Germany

Printed by Libri Plureos GmbH
in Hamburg, Germany